BRAIN RULES

BONUS MATERIAL

www.brainrules.net

br**a**in
rules

Film featuring John Medina

Take a lively, 45-minute tour of the
12 original Brain Rules for home, work,
and school—from "Exercise boosts
brain power" to "Sleep well, think well."

br**a**in
rules
for

Videos guide you through parenting concepts

John Medina hosts fun videos on speaking
in parentese, the cookie experiment,
dealing with temper tantrums, and more.
Plus, take our parenting quiz.

JOHN MEDINA is a developmental molecular biologist and research consultant. He is an affiliate professor of bioengineering at the University of Washington School of Medicine. He was the founding director of two brain research institutes: the Brain Center for Applied Learning Research, at Seattle Pacific University, and the Talaris Research Institute, a nonprofit organization originally focused on how infants encode and process information. Medina lives in Seattle, Washington, with his wife and two boys.

brin
rules

12 Principles for Surviving and Thriving
at Work, Home and School

JOHN   MEDINA

Pear
Press

Pear Press
P.O. Box 70525
Seattle, WA 98127-0525
U.S.A.

This book may be purchased for educational, business,
or sales promotional use. For information, please visit
www.pearpress.com

SECOND EDITION

Edited by Tracy Cutchlow
Designed by Greg Pearson

Library of Congress Cataloging-in-Publication Data
is available upon request.

ISBN-13: 978-0-9832633-7-1

10 9 8 7 6 5 4 3 2

Printed in the United States of America.

To Joshua and Noah

*Gratitude, my dear boys, for constantly reminding me
that age is not something that matters unless you are cheese.*

contents

Brain Rules

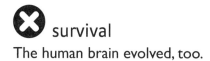 survival
The human brain evolved, too.

 memory
Repeat to remember.

 exercise
Exercise boosts brain power.

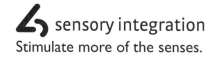 sensory integration
Stimulate more of the senses.

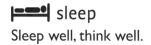 sleep
Sleep well, think well.

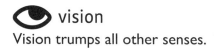 vision
Vision trumps all other senses.

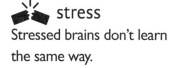 stress
Stressed brains don't learn
the same way.

 music
Study or listen
to boost cognition.

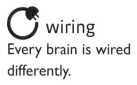 wiring
Every brain is wired
differently.

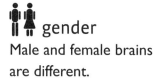 gender
Male and female brains
are different.

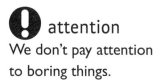 attention
We don't pay attention
to boring things.

 exploration
We are powerful
and natural explorers.

Introduction

GO AHEAD AND MULTIPLY the number 8,388,628 x 2 in your head. Can you do it in a few seconds? There is a young man who can double that number 24 *times* in the space of a few seconds. He gets it right every time. There is a boy who can tell you the precise time of day at any moment, even in his sleep. There is a girl who can correctly determine the exact dimensions of an object 20 feet away. There is a child who at age 6 drew such vivid and complex pictures, some people ranked her version of a galloping horse over one drawn by da Vinci. Yet none of these children have an IQ greater than 70.

The brain is an amazing thing.

Your brain may not be nearly so odd, but it is no less extraordinary. Easily the most sophisticated information-transfer system on Earth, your brain is fully capable of taking the little black squiggles in this book and deriving meaning from them. To accomplish this miracle, your brain sends jolts of electricity crackling through hundreds of miles of wires composed of brain cells so small that thousands of them could fit into the period at the end of this sentence. You accomplish all of this in less time than it takes you to blink. Indeed, you have just done it. What's equally incredible, given our intimate association with it, is this: Most of us have no idea how our brain works.

12 Brain Rules

My goal is to introduce you to 12 things we know about how the brain works. I call these Brain Rules. For each rule, I present the science, introduce you to the researchers behind it, and then offer ideas for how the rule might apply to our daily lives, especially at work and school. The brain is complex, and I am taking only slivers of information from each subject—not comprehensive but, I hope, accessible. Here is a sampling of the ideas you'll encounter:

• We are not used to sitting at a desk for eight hours a day. From an evolutionary perspective, our brains developed while we walked or ran as many as 12 miles a day. The brain still craves this experience. That's why exercise boosts brain power (Brain Rule #2) in sedentary populations like our own. Exercisers outperform couch potatoes in long-term memory, reasoning, attention, and problem-solving tasks.

• As you no doubt have noticed if you've ever sat through a typical PowerPoint presentation, people don't pay attention to boring things (Brain Rule #6). You've got seconds to grab someone's attention and only 10 minutes to keep it. At 9 minutes and 59 seconds, you must do something to regain attention and restart the clock— something emotional and relevant. Also, the brain needs a break. That's why I use stories in this book to make many of my points.

• Ever feel tired about three o'clock in the afternoon? That's because your brain really wants to take a nap. You might be more productive if you did. In one study, a 26-minute nap improved NASA pilots' performance by 34 percent. And whether you get enough rest at night affects your mental agility the next day. Sleep well, think well (Brain Rule #3).

• We'll meet a man who can remember everything he reads after seeing the words just once. Most of us do more forgetting than remembering, of course, and that's why we must repeat to remember (Brain Rule #7). When you understand the brain's rules for memory, you'll see why I want to destroy the notion of homework.

• We'll find out why the terrible twos only look like active rebellion but actually are a child's powerful urge to explore. Babies may not have a lot of knowledge about the world, but they know a whole lot about how to get it. We are powerful and natural explorers (Brain Rule #12). This never leaves us, despite the artificial environments we've built for ourselves.

The grump factor

I am a nice guy, but I am a grumpy scientist. For a study to appear in this book, it has to pass what some of my clients call MGF: the Medina Grump Factor. That means the supporting research for each of my points must first be published in a peer-reviewed journal and then successfully replicated. Many of the studies have been replicated dozens of times. (To stay as reader-friendly as possible, extensive references are not in this book but can be found at *www.brainrules.net/references*.)

No prescriptions

There's a great deal we don't know about the brain. I am a developmental molecular biologist specializing in psychiatric disorders. I have been a private consultant for most of my professional life, working on countless research projects beyond the lab bench. Over and over in my career, I have seen what a distance there is between a gene (one's DNA instructions) and a behavior (how a person actually acts). It's very hard to say with certainty that a specific behavior is caused by a specific gene, or that changing X behavior will produce Y result. Occasionally, I would run across articles and books that made startling claims based on "recent advances" in brain science about how we should teach people and do business. The Mozart Effect comes to mind: the popular idea that listening to classical music makes students better at math. Or the notion that analytical people are "left brain" people and creative people are "right brain" people, and each must be managed accordingly. Sometimes I would panic, wondering if the authors were reading some literature totally off my radar screen. I speak several dialects of brain science, and I knew nothing from those worlds capable of dictating best practices for education and business. In truth, if we ever fully understood how the human brain knew how to pick up a glass of water, it would represent a major achievement. There was no need for me to panic. Brain research still cannot without equivocation tell us how to become better teachers, parents, business leaders, or students. In addition to the ideas you'll find within each chapter, I end each chapter with a few more potential ways to apply the research in our daily lives. But these are not prescriptions. They are hypotheses. If you try them, you will be doing your own little research project to see whether they work for you.

Back to the jungle

What we know about the brain comes from biologists who study brain tissues, experimental psychologists who study behavior, cognitive neuroscientists who study how the first relates to the second, and evolutionary biologists. Though we know precious little about how the brain works, our evolutionary history tells us this: The brain appears to be designed to (1) solve problems (2) related to surviving (3) in an unstable outdoor environment, and (4) to do so in nearly constant motion. I call this the brain's performance envelope.

Each subject in this book—exercise, sleep, stress, wiring, attention, memory, sensory integration, vision, music, gender, and exploration—relates to this performance envelope. We were in motion, getting lots of exercise. Environmental instability led to the extremely flexible way our brains are wired, allowing us to solve problems through exploration. To survive in the great outdoors, we needed to learn from our mistakes. That meant paying attention to certain things at the expense of others, and it meant creating memories in a particular way. Though we have been stuffing them into classrooms and cubicles for decades, our brains actually were built to survive in jungles and grasslands. We have not outgrown this.

Because we don't fully understand how our brains work, we do dumb things. We try to talk on our cell phones and drive at the same time, even though it is literally impossible for our brains to multi-task when it comes to paying attention. We have created high-stress office environments, even though a stressed brain is significantly less productive than a non-stressed brain. Our schools are designed so that most real learning has to occur at home. Taken together, what do the studies in this book show? Mostly this: If you wanted to create an education environment that was directly opposed to what the brain was good at doing, you probably would design something like a classroom. If you wanted to create a business environment that was directly opposed to what the brain was good at doing, you probably would design something like a cubicle. And if you wanted to change things, you might have to tear down both and start over.

Blame it on the fact that brain scientists rarely have a conversation with teachers and business professionals, education majors and accountants, superintendents and CEOs. Unless you have the *Journal of Neuroscience* sitting on your coffee table, you're out of the loop.

This book is meant to get you into the loop.

Survival: Why your brain is so amazing

Brain Rule #1: The human brain evolved, too

When he was 4, my son Noah picked up a stick in our backyard and showed it to me. "Nice stick you have there, young fellow," I said. He replied earnestly, "That's not a stick. That's a sword! Stick 'em up!" I raised my hands to the air. We both laughed. As I went back into the house, I realized my son had just displayed virtually every unique thinking ability a human possesses—one that took several million years to manufacture. And he did so in less than two seconds. Heavy stuff for a 4-year-old. Other animals have powerful cognitive abilities, too, and yet there is something qualitatively different about the way humans think. How and why did our brains evolve this way?

A survival strategy

It all comes down to sex. Our bodies latched on to any genetic adaptation that helped us survive long enough to pass our genes on to the next generation. There's no bigger rule in biology than evolution through natural selection, and the brain is a biological tissue. So it too follows the rule of natural selection.

There are two ways to beat the cruelty of a harsh environment: You can become stronger or you can become smarter. We did the latter. It seems most improbable that such a physically weak species could take over the planet not by adding muscles to our skeletons but by adding neurons to our brains. But we did, and scientists have expended a great deal of effort trying to figure out how. I want to explore four major concepts that not only set the stage for all of the Brain Rules, but also explain how we came to conquer the world.

We can make things up

One trait really does separate us from the gorillas: the ability to use symbolic reasoning. When we see a five-sided geometric shape, we're not stuck perceiving it as a pentagon. We can just as easily perceive the US military headquarters. Or a Chrysler minivan. Our brains can behold a symbolic object as real by itself and yet, simultaneously, also representing something else. That's what my son was doing when he brandished his stick sword. Researcher Judy DeLoache calls it Dual Representational Theory. Stated formally, it describes our ability to attribute characteristics and meanings to things that don't actually possess them. Stated informally, we can make things up that aren't there. We are human because we can fantasize.

We are so good at dual representation, we combine symbols to derive layers of meaning. It gives us the capacity for language, and for writing down that language. It gives us the capacity to reason mathematically. It gives us the capacity for art. Combinations of circles and squares become geometry and Cubist paintings. Combinations of dots and squiggles become music and poetry. There is an unbroken intellectual line between symbolic reasoning and the ability to create culture. And no other creature is capable of doing it.

The all-important human trait of symbolic reasoning helped our species not only survive but thrive. Our evolutionary ancestors didn't have to keep falling into the same quicksand pit if they could tell others about it; even better if they learned to put up warning signs. With words, with language, we could extract a great deal of knowledge about our living situation without always having to experience its harsh lessons directly. It makes sense that once our species evolved to have symbolic reasoning, we kept it. So what was it about our environment that would give a survival advantage to those who could reason symbolically?

We adapted to variation itself

Most of what we know about the intellectual progress of our species is based on evidence of toolmaking. That's not necessarily the most accurate indicator, but it's the best we've got. For the first few million years, the record is not very impressive: We mostly just grabbed rocks and smashed them into things. Scientists, perhaps trying to salvage some of our dignity, called these stones "hand axes." A million years later, we still grabbed "hand axes," but we began to smash them into other rocks, making them more pointed. Now we had sharper rocks. It wasn't much, but it was enough to begin untethering ourselves from a sole reliance on our East African womb, and indeed any other ecological niche. Then things started to get interesting. We created fire and started cooking our food. Eventually, we migrated out of Africa in successive waves, our direct *Homo sapiens* ancestors making the journey as little as 100,000 years ago. Then, 40,000 years ago, something almost unbelievable happened. Our ancestors suddenly took up painting and sculpture, creating fine art and jewelry. This change was both abrupt and profound. Thirty-seven thousand years later, we were making pyramids. Five thousand years after that, rocket fuel.

Many scientists think our growth spurt can be explained by the onset of dual-representation ability. And many think our dual-representation ability—along with physical changes that precipitated it—can be explained by a nasty change in the weather.

Most of human prehistory occurred in junglelike climates: steamy, humid, and in dire need of air-conditioning. This was comfortably predictable. Then the climate changed. Ice cores taken from Greenland show that the climate staggers from being unbearably hot to being sadistically cold. As little as 100,000 years ago, you could be born in a nearly arctic environment but then, mere decades later, be taking off your loincloth to catch the golden rays of the grassland sun. Such instability was bound to have a powerful effect on any creature forced to endure it. Most could not. The rules for survival were changing, and a new class of creatures would start to fill the vacuum created as more and more of their roommates died out.

The change was enough to shake us out of our comfortable trees, but it wasn't violent enough to kill us when we landed. Landing was only the beginning of the hard work, however. Faced with grasslands rather than trees, we were rudely introduced to the idea of "flat." We quickly discovered that our new digs were already occupied. The locals had co-opted the food sources, and most of them were stronger and faster than we were. It is disconcerting to think that we started our evolutionary journey on an unfamiliar horizontal plane with the words "Eat me, I'm prey" taped to our evolutionary butts.

You might suspect that the odds against our survival were great. You would be right. The founding population of our direct ancestors is not thought to have been much larger than 2,000 individuals; some think the group was as small as a few hundred. How, then, did we go from such a wobbly, fragile minority population to a staggering tide of humanity seven billion strong and growing?

There is only one way, according to Richard Potts, director of the Human Origins Program at the Smithsonian's National Museum of Natural History. We gave up on stability. We began not to care about consistency within a given habitat, because consistency wasn't an option. We adapted to variation itself. Those unable to rapidly solve new problems or learn from mistakes didn't survive long enough to pass on their genes. The net effect of this evolution was that rather than becoming stronger, we became smarter. It was a brilliant strategy. We went on to conquer other ecological niches in Africa. Then we took over the world.

Potts's theory predicts some fairly simple things about human learning. It predicts interactions between two powerful features of the brain: a database in which to store a fund of knowledge, and the ability to improvise off that database. One allows us to know when we've made mistakes. The other allows us to learn from them. Both give us the ability to add new information under rapidly changing conditions. And both are relevant to the way we design classrooms and cubicles. We'll uncover more about this database in the Memory chapter.

Bigger and bigger brains

Adapting to variation provides a context for symbolic reasoning, but it hardly explains our unique ability to invent calculus and write romance novels. After all, many animals create a database of knowledge, and many of them make tools, which they use creatively. Still, it is not as if chimpanzees write symphonies badly and we write them well. Chimps can't write them at all, and we can write ones that make people spend their life savings on subscriptions to the New York Philharmonic. There must have been something else in our evolutionary history that gave rise to unique human thinking.

One of the random genetic mutations that gave us an adaptive advantage involved walking upright on two legs. Because the trees were gone or going, we needed to travel increasingly long distances between food sources. Walking on two legs instead of four both freed up our hands and used fewer calories. It was energy-efficient. Our ancestral bodies used the energy surplus not to pump up our muscles but to pump up our minds.

This led to the masterpiece of evolution, the region that distinguishes humans from all other creatures. It is a specialized area of the frontal lobe, just behind the forehead, called the prefrontal cortex. What does the prefrontal cortex do? We got our first hints from a man named Phineas Gage, who suffered the most famous occupational injury in the history of brain science.

Gage was a popular foreman of a railroad construction crew. He was funny, clever, hardworking, and responsible, the kind of guy any father would be proud to call "son-in-law." On September 13, 1848, he set an explosives charge in the hole of a rock using a tamping iron, a three-foot rod about an inch in diameter. The charge blew the rod into Gage's head. It entered just under the eye and destroyed most of his prefrontal cortex. Miraculously, Gage survived. But he became tactless, impulsive, and profane. He left his family and wandered aimlessly from job to job. His friends said he was no longer Gage.

When damage occurs to a specific brain region, we know that any observed behavioral abnormality must in some way be linked to that region's function. I describe several such cases throughout the book for this reason. Gage's case was the first real evidence that the prefrontal cortex governs several uniquely human cognitive talents, called "executive functions": solving problems, maintaining attention, and inhibiting emotional impulses. In short, this region controls many of the behaviors that separate us from other animals (and from teenagers).

Three brains in one

The prefrontal cortex, however, is only the newest addition to the brain. Three brains are tucked inside your head, and parts of their structure took millions of years to design. Your most ancient neural structure is the brain stem, or "lizard brain." This rather insulting label reflects the fact that the brain stem functions the same way in you as in a Gila monster. The brain stem controls most of your body's housekeeping chores: breathing, heart rate, sleeping, waking. Lively as Las Vegas, these neurons are always active, keeping your brain buzzing along whether you're napping or wide awake.

Sitting atop your brain stem is your "mammalian brain." It appears in you the same way it does in many mammals, such as house cats, which is how it got its name. It has more to do with your animal survival than with your human potential. Most of its functions involve what some researchers call the "four Fs": fighting, feeding, fleeing, and … reproductive behavior. Several parts of the mammalian brain play a large role in the Brain Rules.

The amygdala allows you to feel rage. Or fear. Or pleasure. Or memories of past experiences of rage, fear, or pleasure. The amygdala is responsible for both the creation of emotions and the memories they generate. We'll explore the powerful effects of emotions, and how to harness them, in the Attention chapter.

The hippocampus converts your short-term memories into longer-term forms. The Memory chapter covers the surprising way that happens, and the key to remembering.

The thalamus is one of the most active, well-connected parts of the brain—a control tower for the senses. Sitting squarely in the center of your brain, it processes and routes signals sent from nearly every corner of your sensory universe. We'll return to this bizarre, complex process in the Sensory Integration chapter.

Folded atop all of this is your "human brain," a layer called the cortex. Unfolded, this layer would be about the size of a baby blanket, with a thickness ranging from that of blotting paper to that of heavy-duty cardboard. It is in deep electrical communication with the interior. Neurons spark to life, then suddenly blink off, then fire again. Complex circuits of electrical information crackle in coordinated, repeated patterns, racing to communicate their information along large neural highways that branch suddenly into thousands of exits. As we'll see in the Wiring chapter, these branches are different in every single one of us. Each region of the cortex is highly specialized, with sections for speech, for vision, for memory.

You wouldn't know all this just by looking at the brain. The cortex looks homogenous, somewhat like the shell of a walnut, which fooled anatomists for hundreds of years. Then World War I happened. It was the first major conflict where medical advances allowed large numbers of combatants to survive shrapnel injuries. Some of these injuries penetrated only to the periphery of the brain, destroying tiny regions of the cortex while leaving everything else intact. Enough soldiers were hurt that scientists could study in detail the injuries and the truly strange behaviors that resulted. Eventually, scientists were able to make a complete structure–function map of the brain. They were able to see that the brain had, over eons, become three.

Scientists found that as our brains evolved, our heads did, too: They were getting bigger all the time. But the pelvis—and birth canal—can be only so wide, which is bonkers if you are giving birth to children with larger and larger heads. A lot of mothers and babies died on the way to reaching an anatomical compromise. Human pregnancies are still remarkably risky without modern medical intervention. The solution? Give birth while the baby's head is small enough to fit through the birth canal. The problem? You create childhood. Most mammals reach adulthood within months. Our long childhood gave the brain time to finish its developmental programs outside the womb. It also created a creature vulnerable to predators for years and not reproductively fit for more than a decade. That's an eternity when you live in the great outdoors, as we did for eons.

But the trade-off was worth it. A child was fully capable of learning just about anything and, at least for the first few years, not good for doing much else. This created the concept not only of learner but, for adults, of teacher. Of course, it was no use having babies who took years to grow if the adults were eaten before they could finish their thoughtful parenting. We weaklings needed to outcompete the big boys on their home turf, leaving our new home safer for sex and babies. We decided on a strange strategy. We decided to try to get along with each other.

We cooperated: You scratch my back ...

Trying to fight off a woolly mammoth? Alone, and the fight might look like Bambi vs. Godzilla. Two or three of you together—coordinating behavior and establishing the concept of "teamwork"—and you present a formidable challenge. You can figure out how to compel the mammoth to tumble over a cliff, for one. There is ample evidence that this is exactly what we did.

This changes the rules of the game. We learned to cooperate, which means creating a shared goal that takes into account our allies' interests as well as our own. In order to understand our allies' interests, we must be able to understand others' motivations, including their reward and punishment systems. We need to know where their "itch" is. To do this, we constantly make predictions about other people's mental states. Say we hear news about a couple: *The husband died, and then the wife died.* Our minds start working to infer the mental state of the wife: *The husband died, and then the wife died of grief.*

We create a view, however brief, into the psychological interior of the wife. We have an impression of her mental state, perhaps even knowledge about her relationship with her husband. These inferences are the signature characteristic of something called

Theory of Mind. We activate it all the time. We try to see our entire world in terms of motivations, ascribing motivations to our pets and even to inanimate objects. The skill is useful for selecting a mate, for navigating the day-to-day issues surrounding living together, for parenting. Theory of Mind is something humans have like no other creature. It is as close to mind reading as we are likely to get.

This ability to peer inside somebody's mental life and make predictions takes a tremendous amount of intelligence and, not surprisingly, brain activity. Knowing where to find fruit in the jungle is cognitive child's play compared with predicting and manipulating other people within a group setting. Many researchers believe a direct line exists between the acquisition of this skill and our intellectual dominance of the planet.

When we try to predict another person's mental state, we have physically very little to go on. Signs do not appear above a person's head, flashing in bold letters his or her motivations. We are forced to detect something that is not physically obvious at all, such as fear, shame, greed, or loyalty. This talent is so automatic, we hardly know when we do it. We began doing it in every domain. Remember dual representation: the stick and the thing that the stick represents? Our intellectual prowess, from language to mathematics to art, may have come from the powerful need to predict our neighbor's psychological interiors. As I said, your brain is amazing.

Why did I want to spend time walking you through the brain's survival strategies? Because they aren't just part of our species' ancient history. They give us real insight into how humans acquire knowledge. We improvise off a database, thinking symbolically about our world. We are predisposed to social cooperation, which requires constantly reading other people. Along with the performance envelope, these concepts determine at the most fundamental level how our brains work.

Now that you've gotten the gist of things, let's dive into the details.

Brain Rule #1
The human brain evolved, too.

* The brain appears to be designed to (1) solve problems (2) related to surviving (3) in an unstable outdoor environment, and (4) to do so in nearly constant motion.

* We started with a "lizard brain" to keep us breathing, then added a brain like a cat's, and then topped those with the thin layer known as the cortex—the third, and powerful, "human" brain.

* We adapted to change itself, after we were forced from the trees to the savannah when climate swings disrupted our food supply.

* Going from four legs to two to walk on the savannah freed up energy to develop a complex brain.

* Symbolic reasoning is a uniquely human talent. It may have arisen from our need to understand one another's intentions and motivations. This allowed us to coordinate within a group, which is how we took over the Earth.

exercise

Exercise boosts brain power.

IF THE CAMERAS WEREN'T rolling and the media abuzz with live reports, it is possible nobody would have believed the following story:

A man had been handcuffed, shackled, and thrown into California's Long Beach Harbor, where he was quickly fastened to a floating cable. The cable had been attached at the other end to 70 boats, bobbing up and down in the harbor, each carrying a single person. Battling strong winds and currents, the man then swam, towing all 70 boats (and passengers) behind him, traveling 1½ miles from Queensway Bridge. The man, Jack LaLanne, was celebrating his birthday.

He had just turned 70 years old.

Jack LaLanne, born in 1914, has been called the godfather of the American fitness movement. He starred in one of the longest-running exercise programs produced for commercial television. A prolific inventor, LaLanne designed the first leg-extension machines, the first cable-fastened pulleys, and the first weight selectors, all now standard issue in the modern gym. He is credited with inventing an exercise that supposedly bears his name, the Jumping Jack. LaLanne lived to the age of 96. But even these feats are probably not the most interesting aspect of this famed bodybuilder's story.

If you watch him during an interview late in his life, your biggest impression will be not the strength of his muscles but the strength of his *mind*. LaLanne is mentally alert. His sense of humor is both lightning fast and improvisatory. "I tell people I can't afford to die. It will wreck my image!" he joked to Larry King. He once railed: "Do you know how many calories are in butter and cheese and ice cream? Would you get your *dog* up in the morning for a cup of coffee and a donut?" (He claims he hasn't had dessert since 1929.) He has the energy of an athlete in his 20s, and he is possessed of an impressive intellectual vigor.

So it's hard not to ask, "Is there a relationship between exercise and mental alertness?" The answer, it turns out, is yes.

Survival of the fittest

Though a great deal of our evolutionary history remains shrouded in controversy, the one fact that every paleoanthropologist on the planet accepts can be summarized in two words:

We moved.

A lot. As soon as our *Homo erectus* ancestors evolved, about 2 million years ago, they started moving out of town. Our direct ancestors, *Homo sapiens,* rapidly did the same thing. Because bountiful rainforests began to shrink, collapsing the local food supply, our ancestors were forced to wander an increasingly dry landscape looking for more trees to scamper up and dine on. Instead of moving up, down, and across complex arboreal environments, which required a lot of dexterity, we began walking back and forth across arid savannahs, which required a lot of stamina. *Homo sapiens* started in Africa and then took a victory lap around the rest of the world. The speed of the migration is uncertain; the number changes as we find new physical evidence of habitation and as we're better able to isolate and characterize ancient DNA. Anthropologists can say that our ancestors moved fast and they moved far. Males may have walked and run 10 to 20 kilometers a day, says anthropologist Richard Wrangham. The estimate for females is half that. Up to 12 miles: That's the amount of ground scientists estimate we covered *every day*. That means our fancy brains developed not while we were lounging around but while we were exercising.

Regardless of its exact speed, our ancestors' migration is an impressive feat. This was no casual stroll on groomed trails. Early travelers had to contend with fires and floods, insurmountable mountain ranges, foot-rotting jungles, and moisture-sucking deserts. They had no GPS to reassure them, no real tools to speak of. Eventually they made oceangoing boats, without the benefit of wheels or metallurgy, and then traveled up and down the Pacific with only the crudest navigational skills. Our ancestors constantly encountered new food sources, new predators, new physical dangers. Along the way they routinely suffered injuries, experienced strange illnesses, and delivered and nurtured offspring, all without the benefit of textbooks or modern medicine. Given our relative wimpiness in the animal kingdom (we don't even have enough body hair to survive a mildly chilly night), what these data tell us is that we grew up in top physical shape, or we didn't grow up at all. These data also tell us the human brain became the most powerful in the world under conditions where motion was a constant presence.

If our unique cognitive skills were forged in the furnace of physical activity, is it possible that physical activity still influences our cognitive skills? Are the cognitive abilities of someone in good physical condition different from those of someone in poor physical condition? And what if someone in poor physical condition were whipped into shape? Those are scientifically testable questions. The answers are directly related to why Jack LaLanne can still crack jokes about eating dessert. *In his nineties.*

Will you age like Jim or like Frank?

Scientists discovered the beneficial effects of exercise on the brain by looking at aging populations. Years ago while watching television, I came across a documentary on American nursing homes. It showed people in wheelchairs, many in their mid- to late 80s, lining the halls of a dimly lit facility, just sitting around, seemingly waiting to die. One was named Jim. His eyes seemed vacant, lonely, friendless. He could cry at the drop of a hat but otherwise spent the last years of his life mostly staring off into space. I switched channels. I stumbled upon a very young-looking Mike Wallace. The journalist was interviewing architect Frank Lloyd Wright, in his late 80s. I was about to hear a most riveting conversation.

"When I walk into St. Patrick's Cathedral ... here in New York City, I am enveloped in a feeling of reverence," said Wallace, tapping his cigarette.

The old man eyed Wallace. "Sure it isn't an inferiority complex?"

"Just because the building is big and I'm small, you mean?"

"Yes."

"I think not."

"I hope not."

"You feel nothing when you go into St. Patrick's?"

"Regret," Wright said without a moment's pause, "because it isn't the thing that really represents the spirit of independence and the sovereignty of the individual which I feel should be represented in our edifices devoted to culture."

I was dumbfounded by the dexterity of Wright's response. In the space of a few moments, one could detect the clarity of his mind, his unshakable vision, his willingness to think outside the box. The rest of the interview was just as compelling, as was the rest of Wright's life. He completed the designs for the Guggenheim Museum, his last work, in 1957, when he was 90 years old. But I also was dumbfounded by something else. As I contemplated Wright's answers,

I remembered Jim from the nursing home. *He was the same age as Wright.* In fact, most of the residents were. I was beholding two types of aging. Jim and Frank lived in roughly the same period of time. But one mind had almost completely withered, seemingly battered and broken by the aging process, while the other mind remained as incandescent as a light bulb.

What was the difference in the aging process between men like Jim and the famous architect? This question has intrigued the research community for a long time. Attempts to explain these differences led to many important discoveries. I have grouped them as answers to six questions.

1) Is there one factor that predicts how well you will age?

When research on aging began, this question was a tough one to answer. Researchers found many variables, stemming from both nature and nurture, that contributed to someone's ability to age gracefully. That's why the scientific community was both intrigued and cautious when a group of researchers uncovered a powerful environmental influence. One of the greatest predictors of successful aging, they found, is the presence or absence of a sedentary lifestyle.

Put simply, if you are a couch potato, you are more likely to age like Jim, if you make it to your 80s at all. If you have an active lifestyle, you are more likely to age like Frank Lloyd Wright—and much more likely to make it to your 90s. The chief reason for the longer life is that exercise improves cardiovascular fitness, which in turn reduces the risk for diseases such as heart attacks and stroke. But researchers wondered why the people who were aging well also seemed to be more mentally alert. This led to an obvious second question.

2) Were they more mentally alert?

Just about every mental test possible was tried. No matter how it was measured, the answer was consistently yes: A lifetime of exercise results in a sometimes astonishing elevation in cognitive performance, compared with those who are sedentary. Exercisers outperform couch potatoes in tests that measure long-term memory, reasoning, attention, and problem-solving skill. The same is true of fluid-intelligence tasks, which test the ability to reason quickly, think abstractly, and improvise off previously learned material in order to solve a new problem. Essentially, exercise improves a whole host of abilities prized in the classroom and at work.

What about people who aren't elderly? Here, the number of studies done thins out. But in one case, researchers looked at more than 10,000 British civil servants between the ages of 35 and 55, grading their activity levels as low, medium, or high. Those with low levels of physical activity were more likely to have poor cognitive performance. Fluid intelligence, the type that requires improvisatory problem-solving skills, was particularly hurt by a sedentary lifestyle.

Not every cognitive ability is improved by exercise, however. Short-term memory, for example, and certain types of reaction times appear to be unrelated to physical activity. And, while nearly everybody shows some improvement, the degree varies quite a bit among individuals. It's one thing to look at a group of people and note, as early studies did, that those who exercise are also smarter. It's another thing to prove that exercise is the direct cause of the benefits. A more intrusive set of experiments needed to be done to answer the next question.

3) Can you turn Jim into Frank?

Like producers of a makeover show, researchers found a group of elderly couch potatoes, measured their brain power, exercised them, and then reexamined their brain power. The researchers consistently found that all kinds of mental abilities began to come back online— after as little as four months of aerobic exercise. A different study looked at school-age children. Children jogged for 30 minutes two or three times a week. After 12 weeks, their cognitive performance had improved significantly compared with prejogging levels. When the exercise program was withdrawn, the scores plummeted back to their preexperiment levels. Scientists had found a direct link. Within limits, it does appear that exercise can turn Jim into Frank, or at least turn Jim into a sharper version of himself.

As the effects of exercise on cognition became increasingly clear, scientists asked the question dearest to the couch-potato cohort:

4) What type of exercise must you do, and how much?

After years of investigating aging populations, researchers' answer to the question of how much is *not much*. If all you do is walk several times a week, your brain will benefit. Even couch potatoes who fidget show increased benefit over those who do not fidget. The body seems to be clamoring to get back to its active Serengeti roots. Any nod toward this evolutionary history, be it ever so small, is met with a cognitive war whoop. In the laboratory, the gold standard appears to be aerobic exercise, 30 minutes at a clip, two or three times a week. Add a strengthening regimen and you get even more cognitive benefit. Individual results vary, of course, and exercising too intensely, to exhaustion, can hurt cognition. One should consult a physician before embarking on an exercise program. The data merely point to the fact that one should embark. Exercise, as millions of years traipsing around the globe tell us, is good for the brain. Just how good took everyone by surprise, as they delved into the next question.

5) Can exercise treat dementia or depression?

Given the robust effect of exercise on typical cognitive performance, researchers wanted to know if it would have an effect on atypical performance. What about diseases such as age-related dementia and its more thoroughly investigated cousin, Alzheimer's disease? What about affective (mood) disorders such as depression? Researchers looked at both prevention and intervention. With experiments reproduced all over the world, enrolling thousands of people, often studied for decades, the results are clear. Your lifetime risk for general dementia is literally cut in half if you participate in physical activity. Aerobic exercise seems to be the key. With Alzheimer's, the effect is even greater: Such exercise reduces your odds of getting the disease by more than 60 percent.

How much exercise? Once again, a little goes a long way. The researchers showed you have to participate in some form of exercise just twice a week to get the benefit. Bump it up to a 20-minute walk each day, and you can cut your risk of having a stroke—one of the leading causes of mental disability in the elderly—by 57 percent.

Dr. Steven Blair, the man most responsible for stimulating this line of inquiry, did not start his career wanting to be a scientist. He wanted to be an athletics coach. Surely he was inspired by his own football coach in high school, Gene Bissell. Bissell once forfeited a winning game. He realized after the game that an official had missed a call, and he insisted that his team be penalized. Young Steven never forgot the incident. But Bissell encouraged Blair to continue his interest in research, and Blair went on to write a seminal paper on fitness and mortality. The study stands as a landmark example of how to do work with rigor and integrity in this field. His analysis inspired other investigators. What about using exercise not only as prevention, they asked, but as intervention, to treat mental disorders such as depression and anxiety? That turned out to be a good line of questioning.

A growing body of work now suggests that physical activity can powerfully affect the course of both diseases. We think it's because exercise regulates the release of most of the biochemicals associated with maintaining mental health. In one experiment on depression, rigorous exercise was substituted for antidepressant medication. Even when compared to medicated controls, the treatment outcomes were astonishingly successful. For both depression and anxiety, exercise is beneficial immediately and over the long term. It is equally effective for men and women. The longer the person exercises, the greater the effect. Although exercise is not a substitute for psychiatric treatment (which usually involves therapy along with medication), the role of exercise on mood is so pronounced that many psychiatrists prescribe physical activity as well. It is especially helpful in severe cases and for older people.

In asking what else exercise can do, researchers looked beyond our oldest members to our youngest.

6) Does exercise help kids do better in school?

The number of studies in children is downright microscopic. Still, the data point in a familiar direction. Physically fit children identify visual stimuli much faster than sedentary ones. They appear to concentrate better. Brain-activation studies show that children and adolescents who are fit allocate more cognitive resources to a task and do so for longer periods of time. "Kids pay better attention to their subjects when they've been active," Dr. Antronette Yancey said in an interview with NPR. "Kids are less likely to be disruptive in terms of their classroom behavior when they're active. Kids feel better about themselves, have higher self-esteem, less depression, less anxiety. All of those things can impair academic performance and attentiveness."

Of course, many ingredients make up academic performance. Finding out what those components are—and then which are most important for improving performance—is difficult. But these preliminary findings hint that exercise may be one key ingredient.

An exercise in road building

Why exercise works so well in the brain, at a molecular level, can be illustrated by competitive food eaters—or, less charitably, professional pigs. The crest of the International Federation of Competitive Eating proudly displays the motto In Voro Veritas—literally, "In Gorging, Truth." Like any sporting organization, competitive food eaters have their heroes. The reigning gluttony god is Takeru "Tsunami" Kobayashi. He is the recipient of many eating awards, including the vegetarian dumpling competition (83 dumplings downed in eight minutes), the roasted pork bun competition (100 in 12 minutes), and the hamburger competition (97 in eight minutes). Kobayashi also is a world champion hot-dog eater. One of his few losses was to a 1,089-pound Kodiak bear. In a 2003 Fox-televised special called Man vs. Beast, the mighty Kobayashi consumed only 31 bunless dogs compared with the ursine's 50, all in about 2½ minutes. The Tsunami would not accept defeat. In 2012, Kobayashi ate 60 bunless dogs in that amount of time. But my point isn't about speed.

Like the Tsunami's, the brain's appetite for energy is enormous. The brain gobbles up 20 percent of the body's energy, even though it's only about 2 percent of the body's weight. When the brain is fully working, it uses more energy per unit of tissue weight than a fully exercising quadricep. In fact, the human brain cannot simultaneously activate more than 2 percent of its neurons at any one time. More than this, and the brain's energy supply becomes so quickly exhausted that you will faint.

That energy supply is glucose, a type of sugar that is one of the body's favorite resources. After all of those hot dogs slide down the Tsunami's throat, his stomach's acid and his wormy intestines tear the food apart (not getting much help from the teeth, in his case) and reconfigure it into glucose. Glucose and other metabolic products are absorbed into the bloodstream via the small intestines. The nutrients travel to all parts of the body, where they are deposited into cells, which make up the body's various tissues. The cells seize the sweet stuff like sharks in a feeding frenzy. Cellular chemicals greedily tear apart the molecular structure of glucose to extract its sugary energy.

This energy extraction is so violent that atoms are literally ripped asunder in the process. As in any manufacturing process, such fierce activity generates a fair amount of toxic waste. In the case of food, this waste consists of a nasty pile of excess electrons shredded from the atoms in the glucose molecules. Left alone, these electrons slam into other molecules within the cell, transforming them into some of the most toxic substances known to humankind. They are called free radicals. If not quickly corralled, they will wreck havoc on the innards of a cell and, cumulatively, on the rest of the body. These electrons are fully capable, for example, of causing mutations in your DNA.

The reason you don't die of electron overdose is that the atmosphere is full of breathable oxygen. The main function of oxygen is to act like an efficient electron-absorbing sponge. At the same time the blood is delivering glucose to your tissues, it is also carrying these oxygen sponges. Any excess electrons are absorbed by the oxygen and, after a bit of molecular alchemy, are transformed into equally hazardous—but now fully transportable—carbon dioxide. The blood is carried back to your lungs, where the carbon dioxide leaves the blood and you exhale it. So whether you are a competitive eater or a typical one, the oxygen-rich air you inhale keeps the food you eat from killing you. How important is oxygen? The three requirements for human life are food, drink, and fresh air. But their effects on survival have very different timelines. You can live for 30 days or so without food, and you can go for a week or so without drinking water. Your brain, however, is so active that it cannot go without oxygen for more than five minutes without risking serious and permanent damage. When the blood can't deliver enough oxygen sponges, toxic electrons overaccumulate.

Getting energy into tissues and getting toxic electrons out are essentially matters of access. That's why blood—acting as both waitstaff and hazmat team—has to be everywhere inside you. Any tissue without enough blood supply is going to starve to death, your brain included. More access to blood is better. And even in a healthy brain, the blood's delivery system can be improved.

That's where exercise comes in.

It reminds me of a seemingly mundane little insight that literally changed the history of the world. John Loudon McAdam, a Scottish engineer living in England in the early 1800s, noticed the difficulty people had trying to move goods and supplies over hole-filled, often muddy, frequently impassable dirt roads. He had the splendid idea of raising the level of the road using layers of rock and gravel. This immediately made the roads less muddy and more stable. As county after county adopted his process, now called macadamization, people instantly got more dependable access to one another's goods and services. Offshoots from the main roads sprang up. Pretty soon entire countrysides had access to far-flung points using stable arteries of transportation. Trade grew. People got richer. By changing the way things moved, McAdam changed the way we lived.

What does this have to do with exercise? McAdam's central notion wasn't to improve goods and services, but to improve *access* to goods and services. You can do the same for your brain by increasing the roads in your body, namely your blood vessels, through exercise. Exercise does not provide the oxygen and the food. It provides your body greater *access* to the oxygen and the food.

How this works is easy to understand. When you exercise, you increase blood flow across the tissues of your body. Blood flow improves because exercise stimulates the blood vessels to create a powerful, flow-regulating molecule called nitric oxide. As the flow improves, the body makes new blood vessels, which penetrate deeper and deeper into the tissues of the body. This allows more access to the bloodstream's goods and services, which include food distribution and waste disposal. The more you exercise, the more tissues you can feed and the more toxic waste you can remove. This happens all over the body. That's why exercise improves the performance of most human functions. You stabilize existing transportation structures and add new ones, just like McAdam's roads. All of a sudden, you are becoming healthier.

The same happens in the human brain. Imaging studies have shown that exercise increases blood volume in a region of the brain called the dentate gyrus. That's a big deal. The dentate gyrus is a vital constituent of the hippocampus, a region deeply involved in memory formation. This blood-flow increase, likely the result of new capillaries, allows more brain cells greater access to the blood's waitstaff and hazmat team.

Another brain-specific effect of exercise is becoming clear. Early studies indicate that exercise also aids in the development of healthy tissue by stimulating one of the brain's most powerful growth factors, BDNF. That stands for brain-derived neurotrophic factor. "I call it Miracle-Gro, brain fertilizer," says Harvard psychiatrist John Ratey. "It keeps [existing] neurons young and healthy, and makes them more ready to connect with one another. It also encourages neurogenesis— the creation of new cells." The cells most sensitive to this are in the hippocampus, inside the very regions deeply involved in human cognition. Exercise increases the level of usable BDNF inside those cells. Most researchers believe this uptick also buffers against the negative molecular effects of stress, which in turn may improve memory formation. We'll have more to say about this interaction in the Stress chapter.

Redefining normal

All of the evidence points in one direction: Physical activity is cognitive candy. Civilization, while giving us such seemingly forward advances as modern medicine and spatulas, also has had a nasty side effect. It gives us more opportunities to sit on our butts. Whether learning or working, we gradually quit exercising the way our ancestors did. Recall that our evolutionary ancestors were used to walking up to 12 miles *per day*. This means that our brains were supported for most of our evolutionary history by Olympic-caliber bodies. We were not sitting in a classroom for eight hours at a stretch. We were not sitting in a cubicle for eight hours at a stretch. If we sat around the Serengeti for eight hours—heck, for eight *minutes*—we were usually somebody's lunch. We haven't had millions of years to adapt to our sedentary lifestyle. That lifestyle has hurt both our physical and mental health. There is no question we are living in an epidemic of fatness, a point I will not belabor here. The benefits of exercise seem nearly endless because its impact is systemwide, affecting most physiological systems. Exercise makes your muscles and bones stronger, improving your strength and balance. It helps regulate your appetite, reduces your risk for more than a dozen types of cancer, improves the immune system, changes your blood lipid profile, and buffers against the toxic effects of stress (see the Stress chapter). By enriching your cardiovascular system, exercise decreases your risk for heart disease, stroke, and diabetes. When combined with the intellectual benefits exercise appears to offer, we have in our hands as close to a magic bullet for improving human health as exists in modern medicine. So I am convinced that integrating exercise into those eight hours at work or school will only make us *normal*.

All we have to do is move.

More ideas

I can think of a few simple ways to harness the effects of exercise in the practical worlds of education and business.

Recess twice a day

Because of the increased reliance on test scores for school survival, many districts across the nation are getting rid of physical education and recess. Given the powerful cognitive effects of physical activity, this makes no sense. Dr. Yancey described a real-world test: "They took time away from academic subjects for physical education ... and found that, across the board, [adding exercise] did not hurt the kids' performance on the academic tests. [When] trained teachers provided the physical education, the children actually did better on language, reading, and the basic battery of tests."

Cutting off physical exercise—the very activity most likely to promote cognitive performance—to do better on a test score is like trying to gain weight by starving yourself. A smarter approach would be to insert more, not less, exercise into the daily curriculum. They might even reintroduce the notion of school uniforms. Of what would the new apparel consist? Simply gym clothes, worn all day long. If your children's school isn't on board, consider how you could help your kids get 20 to 30 minutes each morning for aerobic exercise; and 20 to 30 minutes each afternoon for strengthening exercises. Most studies show a benefit from exercising only two or three times a week.

You could apply the same idea at work, taking morning and afternoon breaks for exercise. Conduct meetings while you walk, whether in the office or outside. You just might see a boost in problem solving and creativity.

Treadmills and bikes in classrooms and cubicles

Remember the experiment showing that when children aerobically exercised, their brains worked better, and when the exercise stopped, the cognitive gain soon plummeted? These results suggested to the researchers that one's level of fitness is not as important as a steady increase in oxygen to the brain. Otherwise, the improved mental sharpness would not have fallen off so rapidly. So they did another experiment. They administered supplemental oxygen to young healthy adults, and they found a cognitive improvement similar to that of exercise. This suggests an interesting idea to try in a classroom. (Don't worry, it doesn't involve oxygen doping.)

What if, during a lesson, the children were not sitting at desks but walking on treadmills or riding stationary bikes? Students might study English while peddling comfortably on a bike that accommodates a desk. Workers could easily do the same, composing email while walking on a treadmill at one to two miles per hour. This idea would harness the advantage of increasing the oxygen supply and at the same time harvest all the other advantages of regular exercise.

The idea of integrating exercise into the workday or school day may sound foreign, but it's not difficult. I put a treadmill in my own office, and I now take regular breaks filled not with coffee but with exercise. I constructed a small structure upon which my laptop fits so that I can write while I walk. At first, it was difficult to adapt to such a strange hybrid activity. It took a whopping 15 minutes to become fully functional typing on my laptop while walking 1.8 miles per hour.

Office workers can sometimes choose their own desk setups, integrating exercise on an individual basis. But businesses have compelling reasons to incorporate such radical ideas into company policy as well. Business leaders already know that if employees exercised regularly, it would reduce health-care costs. There's no question that halving someone's lifetime risk of a debilitating stroke or Alzheimer's disease is a wonderfully humanitarian thing to do. But exercise also could boost the collective brain power of an organization. Fit employees are more capable than sedentary employees of mobilizing their God-given IQs. For companies whose competitiveness rests on creative intellectual horsepower, such mobilization could mean a strategic advantage. In the laboratory, regular exercise improves problem-solving abilities, fluid intelligence, and even memory— sometimes dramatically so. It's worth finding out whether the same is true in business settings, too.

Brain Rule #2
Exercise boosts brain power.

- Our brains were built for walking—12 miles a day!

- To improve your thinking skills, *move.*

- Exercise gets blood to your brain, bringing it glucose for energy and oxygen to soak up the toxic electrons that are left over. It also stimulates the protein that keeps neurons connecting.

- Aerobic exercise just twice a week halves your risk of general dementia. It cuts your risk of Alzheimer's by 60 percent.

Get illustrations, audio, video, and more
at www.brainrules.net

sleep

Sleep well, think well.

IT'S NOT THE MOST comfortable way to raise funds for a major American charity. In 1959, New York disk jockey Peter Tripp decided that he would stay awake for 200 straight hours. He got into a glass booth in the most visible place possible in New York—Times Square—and rigged up the radio so that he could broadcast his show. He even allowed scientists (and, wisely, medical professionals) to observe and measure his behavior as he descended into sleeplessness. One of those scientists was famed sleep researcher William Dement. For the first 72 hours, everything seemed fine with Tripp. He gave his normal three-hour show with humor and professional aplomb. Then things changed. Tripp became rude and offensive to the people around him. Hallucinations set in. The researchers testing his cognitive skills halfway through found he could no longer complete certain mental skill tests. At the 120-hour mark—five days in—Tripp showed real signs of mental impairment, which would only worsen with time. Dement described Tripp's behavior toward the end of the adventure: "The disk jockey developed an acute paranoid psychosis during the nighttime hours, accompanied at times by auditory hallucination. He believed that unknown adversaries were attempting to slip drugs into his food and beverages in order to put him to sleep." At the 200-hour mark—more than eight days—Tripp was done. Presumably, he went to bed and stayed there for a long time.

Some unfortunate souls don't have the luxury of experimenting with sleep deprivation. They become suddenly and permanently incapable of ever going to sleep again. Only about 20 families in the world suffer from Fatal Familial Insomnia, making it one of the rarest human genetic disorders that exists. That rarity is a blessing, because the disease follows a course straight through mental-health hell. In middle to late adulthood, the person begins to experience fevers, tremors, and profuse sweating. As the insomnia becomes permanent, these symptoms are accompanied by increasingly uncontrollable muscular jerks and tics. The person soon experiences crushing feelings of depression and anxiety. He or she becomes psychotic. Finally, mercifully, the patient slips into a coma and dies.

So we know bad things happen when we don't sleep. The puzzle is that, from an evolutionary standpoint, bad things also could happen when we do sleep. Because the body goes into a human version of micro-hibernation, sleep makes us exquisitely vulnerable to predators. Indeed, deliberately going off to dreamland unprotected in the middle of a bunch of hostile hunters (such as leopards, our evolutionary roommates in eastern Africa) seems like a plan dreamed up by our worst enemies. There must be something terribly important we need to accomplish during sleep if we are willing to take such risks in order to get it. Exactly what is it that is so darned important?

To begin to understand *why* we spend a walloping one-third of our time on this planet sleeping, let's peer in on what the brain is doing while we sleep.

You call this rest?

If you ever get a chance to listen in on someone's brain while its owner is slumbering, you'll have to get over your disbelief. The brain does not appear to be asleep at all. Rather, it is almost unbelievably active during "rest," with legions of neurons crackling electrical commands to one another in constantly shifting, extremely active patterns. In fact, the only time you can observe a real resting period for the brain—where the amount of energy consumed is less than during a similar awake period—is during the phase called non-REM sleep. But that takes up only about 20 percent of the total sleep cycle. This is why researchers early on began to disabuse themselves of the notion that the reason we rest is so that we can rest. When we are asleep, the brain is not resting at all. Even so, most people report that sleep is powerfully restorative, and they point to the fact that if they don't get enough sleep, they don't think as well the next day. That is measurably true, as we shall see shortly. And so we find ourselves in a quandary: Given the amount of energy the brain is using, it seems impossible that you could receive anything approaching mental rest and restoration during sleep.

Two scientists made substantial early contributions to our understanding of what the brain is doing while we sleep. Dement, who studied sleepless Peter Tripp, is a white-haired man with a broad smile who at this writing is in his late 80s. He says pithy things about our slumbering habits, such as "Dreaming permits each and every one of us to be quietly and safely insane every night of our lives." Dement's mentor, a gifted researcher named Nathaniel Kleitman, gave him many of his initial insights. If Dement can be considered the father of sleep research, Kleitman certainly could qualify as its grandfather. An intense Russian man with bushy eyebrows, Kleitman may be best noted for his willingness to experiment not only on himself but also on his children. When it appeared that a colleague of his had discovered rapid eye movement (REM) sleep, Kleitman promptly volunteered his daughter for experimentation, and she just as promptly confirmed the finding. He also persuaded a colleague to live with him underground to see what would happen to their sleep cycles without the influence of light and social cues. Here are some of the things Dement and Kleitman discovered about sleep.

Sleep is a battle

Like soldiers on a battlefield, we have two powerful and opposing drives locked in vicious, biological combat. The armies, each made of legions of brain cells and biochemicals, have very different agendas. Though localized in the head, the theater of operations for these armies engulfs every corner of the body. The war they are waging has some interesting rules. First, these forces are engaged not just during the night, while we sleep, but also during the day, while we are awake. Second, they are doomed to a combat schedule in which each army sequentially wins one battle, then promptly loses the next battle, then quickly wins the next and so on, cycling through this win/loss column every day and every night. Third, neither army ever claims final victory. This incessant engagement is referred to as the "opponent process" model. It results in the waking and sleeping modes all humans cycle through every day (and night) of our lives.

One army is composed of neurons, hormones, and various other chemicals that do everything in their power to keep you awake. This army is called the circadian arousal system (often simply called "process C"). If this army had its way, you would stay up all the time. It is opposed by an equally powerful army, also made of brain cells, hormones, and various chemicals. These combatants do everything in their power to put you to sleep. They are termed the homeostatic sleep drive ("process S"). If this army had its way, you would go to sleep and never wake up. These drives define for us both the amount of sleep we need and the amount of sleep we get. Stated formally, process S maintains the duration and intensity of sleep, while process C determines the tendency and timing of the need to go to sleep.

It is a paradoxical war. The longer one army controls the field, for example, the more likely it is to lose the battle. It's almost as if each army becomes exhausted from having its way and eventually waves a temporary white flag. Indeed, the longer you are awake (the victorious process C doing victory laps around your head), the greater the probability becomes that the circadian arousal system will cede the field to its opponent. You then go to sleep. For most people, this act of capitulation comes after about 16 hours of active consciousness. This will occur, Kleitman found, even if you are living in a cave.

Conversely, the longer you are asleep (the triumphant process S now doing the heady victory laps), the greater the probability becomes that the homeostatic sleep drive will similarly cede the field to *its* opponent, which is, of course, the drive to keep you awake. The result of this surrender is that you wake up. For most people, the length of time prior to capitulation is about half of its opponent's, about eight hours of blissful sleep. And this also will occur even if you are living in a cave.

Such dynamic tension is a normal—even critical—part of our daily lives. In fact, the circadian arousal system and the homeostatic sleep drive are locked in a cycle of victory and surrender so predictable, you can graph it.

In one of Kleitman's most interesting experiments, he and a colleague spent an entire month living 1,300 feet underground in Mammoth Cave in Kentucky. Free of sunlight and daily schedules, Kleitman could find out whether the routines of wakefulness and sleep cycled themselves automatically through the human body. His experiment provided the first real hint that such an automatic device did exist in our bodies. Indeed, we now know that the body possesses a series of internal clocks, all controlled by discrete regions in the brain, providing a regular rhythmic schedule to our waking and sleeping experiences. This is surprisingly similar to the buzzing of a wristwatch's internal quartz crystal. An area of the brain called the suprachiasmatic nucleus appears to contain just such a timing device. Of course, we have not been characterizing these pulsing rhythms as a benign wristwatch. We have been characterizing them as a war. One of Kleitman and Dement's greatest contributions was to show that this nearly automatic rhythm occurs as a result of the continuous conflict between two opposing forces.

Are you a lark, owl, or hummingbird?

Each of us wages this war on a slightly different schedule. The late advice columnist Ann Landers apparently would take her phone off the hook between 1:00 a.m. and 10:00 a.m. Why? This was the time she normally slept. "No one's going to call me," she said, "until I'm ready." The cartoonist Scott Adams, creator of the comic strip *Dilbert*, never would think of starting his day at 10:00 a.m. "I'm quite tuned into my rhythms," he told the authors of *The Body Clock Guide to Better Health*. "I never try to do any creating past noon.... I do the strip from 6:00 to 7:00 a.m." Here we have two creative and well-accomplished professionals, one who starts working just as the other's workday is finished.

About one in 10 of us is like *Dilbert's* Adams. The scientific literature calls such people larks (more palatable than the proper term, "early chronotype"). In general, larks report being most alert around noon and feel most productive at work a few hours before they eat lunch. They don't need an alarm clock, because they invariably get up before the alarm rings—often before 6:00 a.m. Larks cheerfully report their favorite mealtime as breakfast and generally consume much less coffee than non-larks. Getting increasingly drowsy in the early evening, most larks go to bed (or want to go to bed) around 9:00 p.m.

Larks are incomprehensible to the one in 10 humans who lie at the other extreme of the sleep spectrum: "late chronotypes," or owls. In general, owls report being most alert around 6:00 p.m., experiencing their most productive work times in the late evening. They rarely want to go to bed before 3:00 a.m. Owls invariably need an alarm clock to get them up in the morning, with extreme owls requiring multiple alarms to ensure arousal. Indeed, if owls had their druthers, most would not wake up much before 10:00 a.m. Not surprisingly, late chronotypes report their favorite mealtime as dinner, and they would drink gallons of coffee all day long to prop themselves up at work if given the opportunity. If it sounds to you as though owls do not sleep as well as larks in American society, you are right on the money. Indeed, late chronotypes usually accumulate a massive "sleep debt" as they go through life.

Whether lark or owl, researchers think these patterns are detectable in early childhood and burned into genes that govern our sleep/wake cycle. At least one study shows that if Mom or Dad is a lark, half of their kids will be, too. Larks and owls, though, cover only about 20 percent of the population. The rest of us are called hummingbirds. True to the idea of a continuum, some hummingbirds are more owlish, some are more larkish, and some are in between.

Nappin' in the free world

It must have taken some getting used to, if you were a staffer in the socially conservative early 1960s. Lyndon Baines Johnson, 36th president of the United States and leader of the free world, routinely closed the door to his office in the midafternoon and put on his pajamas. He then proceeded to take a 30-minute nap. Rising refreshed, he would then resume his role as commander in chief. Such presidential behavior might seem downright weird. But if you asked a sleep researcher like Dement, his response might surprise you: It was LBJ who was acting normally. The rest of us, who refuse to bring our pajamas to work, are the abnormal ones.

LBJ was responding to something experienced by nearly everyone on the planet. It goes by many names—the midday yawn, the post-lunch dip, the afternoon "sleepies." We'll call it the nap zone, a period of time in the midafternoon when we experience transient sleepiness. It can be nearly impossible to get anything done during this time, and if you attempt to push through, which is what most of us do, you can spend much of your afternoon fighting a gnawing tiredness. It's a fight because the brain really wants to take a nap and doesn't care what its owner is doing. The concept of "siesta," institutionalized in many other cultures, may have come as an explicit reaction to the nap zone.

At first, scientists didn't believe the nap zone existed except as an artifact of sleep deprivation. That has changed. We now know that some people feel it more intensely than others. We know it is not related to a big lunch (although a big lunch, especially one loaded with carbs, can greatly increase its intensity). We also know that when you chart the process S curve and process C curve, you can see that they flatline in the same place—in the afternoon. The biochemical battle reaches a climactic stalemate. An equal tension now exists between the two drives, which extracts a great deal of energy to maintain. Some researchers, though not all, think this equanimity in tension drives the need to nap. Some think that a long sleep at night and a short midday nap represent default human sleep behavior, that it is part of our evolutionary history.

Regardless of the cause, the nap zone matters, because our brains don't work as well during it. If you are a public speaker, you already know it is darn near fatal to give a talk in the midafternoon. The nap zone also is literally fatal: More traffic accidents occur during it than at any other time of the day.

If you embrace the need to nap rather than pushing through, as LBJ found, your brain will work better afterward. One NASA study showed that a 26-minute nap reduced a flight crew's lapses in awareness by 34 percent, compared to a control group who didn't nap. Nappers also saw a 16 percent improvement in reaction times. And their performance stayed consistent throughout the day rather than dropping off at the end of a flight or at night. (The flight crew was given a 40-minute break, it took about six minutes for people to fall asleep, and the average nap lasted 26 minutes.) Another study showed that a 45-minute nap produces a similar boost in cognitive performance, a boost lasting more than six hours. Also, napping for 30 minutes before pulling an all-nighter keeps your mind sharper in the wee hours.

What happens if we don't get enough sleep

Given our understanding of how and when we sleep, you might expect that scientists would have an answer to the question of how much sleep we need. Indeed, they do. The answer is: We don't know. You did not read that wrong. After all of these centuries of experience with sleep, we still don't know how much of the stuff people actually need. Generalizations don't work. When you dig into the data on humans, what you find is not remarkable uniformity but remarkable individuality. To make matters worse, sleep schedules are unbelievably dynamic. They change with age. They change with gender. They change depending upon whether or not you are pregnant, and whether or not you are going through puberty. One must take into account so many variables that it almost feels as though we've asked the wrong question.

So let's invert the query. How much sleep *don't* you need? In other words, what are the numbers that disrupt normal function?

Sleep loss = brain drain

One study showed that a highly successful student can be set up for a precipitous academic fall just by getting less than seven hours of sleep a night. Take an A student used to scoring in the top 10 percent of virtually anything she does. If she gets just under seven hours of sleep on weekdays, and about 40 minutes more on weekends, her scores will begin to match the scores of the bottom 9 percent of individuals who are getting enough sleep. Cumulative losses during the week add up to cumulative deficits during the weekend—and, if not paid for, that sleep debt will be carried into the next week.

Another study followed soldiers responsible for operating complex military hardware. One night's loss of sleep resulted in about a 30 percent loss in overall cognitive skill, with a subsequent drop in performance. Bump that to two nights of sleep loss, and the loss in cognitive skill doubles to 60 percent.

Other studies showed that when sleep was restricted to six hours or less per night for just five nights, cognitive performance matched that of a person suffering from 48 hours of continual sleep deprivation.

What do these data tell us? That some people need at least seven hours of sleep a night. And that some people need at least six hours of sleep a night. On the other hand, you may have heard of people who seem to need only four or five hours of sleep. They are referred to as suffering from "healthy insomnia." Essentially, it comes down to whatever amount of sleep is right for you. When robbed of that, bad things really do happen to your brain.

Sleep loss takes a toll on the body, too—on functions that do not at first blush seem associated with sleep. When people become sleep deprived, for example, their body's ability to utilize the food they are consuming falls by about one-third. The ability to make insulin and to extract energy from the brain's favorite source, glucose, begins to fail miserably. At the same time, you find a marked need to have more of it, because the body's stress hormone levels begin to rise in an increasingly deregulated fashion. If you keep up the behavior, you appear to accelerate parts of the aging process. For example, if healthy 30-year-olds are sleep deprived for six days (averaging, in this study, about four hours of sleep per night), parts of their body chemistry soon revert to that of a 60-year-old. And if they are allowed to recover, it will take them almost a week to get back to their 30-year-old systems.

Taken together, these studies show that sleep loss cripples thinking in just about every way you can measure thinking. Sleep loss hurts attention, executive function, working memory, mood, quantitative skills, logical reasoning ability, general math knowledge. Eventually, sleep loss affects manual dexterity, including fine motor control, and even gross motor movements, such as the ability to walk on a treadmill.

So what can a good night's sleep do for us?

Sleep on it: benefits of a solid night's rest

Dimitri Ivanovich Mendeleyev was your archetypal brilliant-but-mad-looking scientist. Hairy and opinionated, Mendeleyev possessed the lurking countenance of a Rasputin, the haunting eyes of Peter the Great, and the moral flexibility of both. He once threatened to commit suicide if a young lady didn't marry him. She consented, which was quite illegal, because unbeknownst to the poor girl, Mendeleyev was already married. This trespass kept him out of the Russian Academy of Sciences for some time, which in hindsight may have been a bit rash, as Mendeleyev single-handedly systematized the entire science of chemistry. His Periodic Table of the Elements—a way of organizing every atom that had so far been discovered—was so prescient, it allowed room for all of the elements yet to be found and even predicted some of their properties.

But what's most extraordinary is this: Mendeleyev says he came up with the idea in his sleep. Contemplating the nature of the universe while playing solitaire one evening, he nodded off. When he awoke, he knew how all of the atoms in the universe were organized, and he promptly created his famous table. Interestingly, he organized the atoms in repeating groups of seven, just the way you play solitaire.

Mendeleyev is hardly the only scientist who has reported feelings of inspiration after having slept. Is there something to the notion of "Let's sleep on it"? Mountains of data say there is. A healthy night's sleep can indeed boost learning significantly. Sleep scientists debate how we should define learning, and what exactly is improvement.

But there are many examples of the phenomenon. One study stands out in particular.

Students were given a series of math problems and prepped with a method to solve them. The students weren't told there was also an easier "shortcut" way to solve the problems, potentially discoverable while doing the exercise. The question was: Is there any way to jump-start, even speed up, the insight into the shortcut? The answer was yes, *if you allow them to sleep on it.* If you let 12 hours pass after the initial training and ask the students to do more problems, about 20 percent will have discovered the shortcut. But, if in that 12 hours you also allow eight or so hours of regular sleep, that figure triples to about 60 percent. No matter how many times the experiment is run, the sleep group consistently outperforms the non-sleep group about three to one.

Sleep also has been shown to enhance tasks that involve visual texture discrimination (the ability to pick out an object from an ocean of similar-looking objects), motor adaptations (improving movement skills), and motor sequence learning. The type of learning that appears to be most sensitive to sleep improvement is that which involves learning a procedure. Simply disrupt the night's sleep at specific stages and retest in the morning, and you eliminate any overnight learning improvement. Clearly, for specific types of intellectual skill, sleep can be a great friend to learning.

Why we sleep

Consider the following true story of a successfully married, incredibly detail-oriented accountant. Even though dead asleep, he regularly gives financial reports to his wife all night long. Many of these reports come from the day's activities. (Incidentally, if his wife wakes him up—which is often, because his financial broadcasts are loud— the accountant becomes amorous and wants to have sex.) Are we all organizing our previous experiences while we sleep? Could this not only explain all of the other data we have been discussing, but also provide the reason why we sleep?

To answer these questions, we turn to a group of researchers who left a bunch of wires stuck inside a rat's brain—electrodes placed near individual neurons. The rat had just learned to negotiate a maze when it decided to take a nap. The wires were attached to a recording device, which happened to still be on. The device allows scientists to eavesdrop on the brain while it is talking to itself, something like an NSA phone tap. Even in a tiny rat's brain, it is not unusual these days to listen in on the chattering of up to 500 neurons at once as they process information. So what are they all saying?

If you listen in while the rat is acquiring new information, like learning to navigate a maze, you soon will detect something extraordinary. A very discrete "maze-specific" pattern of electrical stimulation begins to emerge. Working something like the old Morse code, a series of neurons begin to crackle in a specifically timed sequence while the mouse is learning. Afterward, the rat will always fire off that same pattern whenever it travels through the maze. It appears to be an electrical representation of the rat's new maze-navigating thought patterns (at least, as many as 500 electrodes can detect).

When the rat goes to sleep, its brain begins to *replay the maze-pattern sequence*. Reminiscent of our accountant, the animal's brain repeats what it learned that day. Always executing the pattern in a specific stage of sleep, the rat repeats it over and over again—and much faster than during the day. The rate is so furious, the sequence is replayed thousands of times. If a mean graduate student decides to wake up the rat during this stage, called slow-wave sleep, something equally extraordinary is observed. The rat has trouble remembering the maze the next day. Quite literally, the rat seems to be consolidating the day's learning the night *after* that learning occurred, and an interruption of that sleep disrupts the learning cycle.

This naturally caused researchers to ask whether the same was true for humans. The answer? Not only do we do such processing, but we do it in a more complex fashion. Like the rat, humans appear to replay certain learning experiences at night, during the slow-wave phase. Unlike the rat, more emotionally charged memories appear to replay at a different stage in the sleep cycle.

These findings represent a bombshell of an idea: Some kind of offline processing is occurring at night. Is it possible that the reason we need to sleep is simply to shut off the exterior world for a while, allowing us to divert more attention to our cognitive interiors? Is it possible that the reason we need to sleep is so that we can learn?

It sounds compelling, but of course the real world of research is much messier. Some findings appear to complicate, if not fully contradict, the idea of offline processing. For example, brain-damaged individuals who lack the ability to sleep in the slow-wave phase nonetheless have normal, even improved, memory. So do individuals whose REM sleep is suppressed by antidepressant medications. Exactly how to reconcile these data with the previous findings is a subject of intense scientific debate. Newer findings in mice suggest that the brain uses the time to clean house, sweeping away the toxic molecules that are a byproduct of the brain doing its thinking. With more time and more research, we'll gain a greater understanding of what the brain is doing as we sleep—and why.

For now, a consistent concept emerges: Sleep is intimately involved in learning. It is observable with large amounts of sleep; it is observable with small amounts of sleep; it is observable all the time. It is time we did a better job of observing its importance in our lives.

More ideas

If businesses and schools took sleep seriously, what would a modern office building look like? A modern school? These are not idle questions. The effects of sleep deprivation are thought to cost US businesses more than $100 billion a year.

Match schedules to chronotypes

Behavioral tests can easily discriminate larks from owls from hummingbirds. Given advances in genetic research, in the future you may need only a blood test to characterize your process C and process S graphs. That means you can determine the hours when you are likely to experience productivity peaks. Twenty percent of the workforce is already at suboptimal productivity in the current nine-to-five model. So here's an obvious idea: Set your schedule—whether college class schedule or work schedule—to match your chronotype.

Businesses could create several work schedules, based on the chronotypes of the employees. They might gain more productivity and a greater quality of life for those unfortunate people who otherwise are doomed to carry a permanent sleep debt. A business of the future takes sleep schedules seriously.

We could do the same in education. Teachers are just as likely to be late chronotypes as their students. Why not put them together? You might increase the competencies of both the teacher and the students. Freed of the nagging consequences of their sleep debts, each might be more fully capable of mobilizing his or her God-given IQ.

Variable schedules also would take advantage of the fact that sleep needs change throughout a person's life. For example, data suggest that students temporarily shift to more of an owl chronotype as they transit through their teenage years. This has led some school districts to start their high-school classes after 9:00 a.m. This may make some sense. Sleep hormones (such as the protein melatonin) are at their maximum levels in the teenage brain. The natural tendency of these kids is to sleep more, especially in the morning. As we age, we tend to get less sleep, and some evidence suggests we need less sleep, too. An employee who starts out with her greatest productivity in one schedule may, as the years go by, keep a similar high level of output simply by switching to a different schedule.

Respect the nap zone

Don't schedule meetings or classes during the time when the process C and process S curves are flatlined. Don't give high-demand presentations or take critical exams anywhere near the collision of these two curves. Can you actually get a nap? That's often easier said than done. College students can perhaps get back to their dorm rooms. Stay-at-home parents might be able to sleep when baby does. Some employees sneak out to their cars.

Even better would be if schools and businesses deliberately planned downshifts during the nap zone. Naps would be accorded the same deference that businesses reluctantly treat lunch, or even potty breaks: a necessary nod to an employee's biological needs. Companies could create a designated space for employees to take one half-hour nap each workday. The advantage would be straightforward. People hired for their intellectual strength would be allowed to keep that strength in tip-top shape. "What other management strategy will improve people's performance 34 percent in just 26 minutes?" said Mark Rosekind, the NASA scientist who conducted that eye-opening research on naps and pilot performance.

Sleep on it

Given the data about a good night's rest, organizations might tackle their most intractable problems by having the entire "solving team" go on a mini-retreat. Once arrived, employees would be presented with the problem and asked to think about solutions. But they would not start coming to conclusions, or even begin sharing ideas with each other, before they had slept about eight hours. When they awoke, would the same increase in problem-solving rates available in the lab also be available to that team? It's worth finding out.

Brain Rule #3
Sleep well, think well.

● The brain is in a constant state of tension between cells and chemicals that try to put you to sleep and cells and chemicals that try to keep you awake.

● The neurons of your brain show vigorous rhythmical activity when you're asleep—perhaps replaying what you learned that day.

● People vary in how much sleep they need and when they prefer to get it, but the biological drive for an afternoon nap is universal.

● Loss of sleep hurts attention, executive function, working memory, mood, quantitative skills, logical reasoning, and even motor dexterity.

stress

Brain Rule #4
Stressed brains don't learn
the same way.

IT IS, BY ANY measure, a thoroughly rotten experiment. Here is this beautiful German shepherd, lying in one corner of a metal box, whimpering. He is receiving painful electric shocks, stimuli that should leave him howling in pain. Oddly enough, the dog could easily get out. The other side of the box is perfectly insulated from shocks, and only a low barrier separates the two sides. Though the dog could jump over to safety when the whim strikes him, the whim doesn't strike him. He just lies down in the corner of the electric side, whimpering with each jarring jolt. He must be physically removed by the experimenter to be relieved of the experience.

What has happened to that dog?

A few days before entering the box, the animal was strapped to a restraining harness rigged with electric wires, inescapably receiving the same painful shock day and night. And at first he didn't just stand there taking it, he *reacted*. He howled in pain. He urinated. He strained mightily against his harness in an increasingly desperate attempt to link some behavior of his with the cessation of the pain. But it was no use. As the hours and even days ticked by, his resistance eventually subsided. Why? The dog began to receive a very clear message: The pain was not going to stop; the shocks were going to be forever. *There was no way out.* Even after the dog had been released from the harness and placed into the metal box with the escape route, he could no longer understand his options. Learning had been shut down.

Those of you familiar with psychology already know I am describing a famous set of experiments begun in the late 1960s by legendary psychologist Martin Seligman. He coined the term "learned helplessness" to describe both the perception of inescapability and its associated cognitive collapse. Many animals behave in a similar fashion when punishment is unavoidable, and that includes humans. Inmates in concentration camps routinely experienced these symptoms in response to their horrid conditions. Some camps gave it the name *Gammel*, derived from the colloquial German word *Gammeln*, which literally means "rotting." Perhaps not surprisingly, Seligman spent the rest of his career studying how humans respond to optimism.

What is so awful about severe, chronic stress that it can cause behavioral changes as devastating as learned helplessness? Why is learning so radically altered? We'll begin with a definition of stress, talk about biological responses, and then move to the relationship between stress and learning. Along the way, we will talk about marriage and parenting, about the workplace, and about the first and only time I ever heard my mother, a fourth-grade teacher, swear. It was her first real encounter with learned helplessness.

What is stress? It depends

Not all stress is the same. Certain types of stress really hurt learning, but some types of stress *boost* learning. Second, it's difficult to detect when someone is experiencing stress. Some people love skydiving for recreation; it's others' worst nightmare. Is jumping out of an airplane inherently stressful? The answer is no, and that highlights the subjective nature of stress.

The body alone isn't of much help in providing a definition, either. There is no unique grouping of physiological responses capable of telling a scientist whether you are experiencing stress. That's because many of the mechanisms that cause you to shrink in horror from a predator are the same mechanisms used when you are having sex—or even while you are consuming your Thanksgiving dinner. To your body, saber-toothed tigers and orgasms and turkey gravy look remarkably similar. An aroused physiological state is characteristic of both stress and pleasure.

So what's a scientist to do? A few years ago, gifted researchers Jeansok Kim and David Diamond came up with a three-part definition that covers many of the bases. In their view, if all three are happening simultaneously, a person is stressed.

A measurable physiological response: There must be an aroused physiological response to the stress, and it must be measurable by an outside party. I saw this the first time my then 18-month-old son encountered a carrot on his plate at dinner. He promptly went ballistic: He screamed and cried and peed in his diaper. His aroused physiological state was immediately measurable by his dad, and probably by anyone else within a half mile of our kitchen table.

A desire to avoid the situation: The stressor must be perceived as aversive—something that, given the choice, you'd rather not experience. It was obvious where my son stood on the matter. Within seconds, he snatched the carrot off his plate and threw it on the floor. Then he deftly got down off his chair and tried to stomp on the predatory vegetable.

A loss of control: The person must not feel in control of the stressor. Like a volume knob on some emotional radio, the more the loss of control, the more severe the stress is perceived to be.

This element of control and its closely related twin, predictability, lie at the heart of learned helplessness. My son reacted as strongly as he did in part because he knew I wanted him to eat the carrot, and he was used to doing what I told him to do. Control was the issue. Despite my picking up the carrot, washing it, then rubbing my tummy while enthusiastically saying "yum, yum," he was having none of it. Or, more important, he wanted to have none of it, and he thought I was going to make him have all of it. Feeling out of control over the carrot equaled out-of-control behavior.

When you find this trinity of components working together, you have the type of stress easily measurable in a laboratory setting. When I talk about stress, I am usually referring to situations like these.

We're built for stress that lasts only seconds

You can feel your body responding to stress: Your pulse races, your blood pressure rises, and you feel a massive release of energy. That's the famous hormone adrenaline at work. This fight-or-flight response is spurred into action by your brain's hypothalamus, that pea-size organ sitting almost in the middle of your head. When your sensory systems detect stress, the hypothalamus signals your adrenal glands to dump buckets of adrenaline into your bloodstream. There's a less famous hormone at work, too—also released by the adrenals, and just as powerful as adrenalin. It's called cortisol. It's the second wave of our defensive reaction to stressors. In small doses, it wipes out most unpleasant aspects of stress, returning us to normalcy.

Why do our bodies need to go through all this trouble? The answer is very simple. Without a flexible, immediately available, highly regulated stress response, we would die. Remember, the brain is the world's most sophisticated survival organ. All of its many com-plexities are built toward a mildly erotic, singularly selfish goal: to live long enough to thrust our genes on to the next generation. Our reactions to stress help us manage the threats that could keep us from procreating.

And what kinds of survival threats did we experience in our evolutionary toddlerhood? Predators would make the top 10 list. So would physical injury. In modern times, a broken leg means a trip to the doctor. In our distant past, a broken leg often meant a death sentence. The day's weather would have been a concern, the day's offering of food another. A lot of very *immediate* needs rise to the surface. Most of the survival issues we faced in our first few million years did not take long to settle. The saber-toothed tiger either ate us or we ran away from it—or a lucky few might stab it, but the whole thing was usually over in moments. Consequently, our stress responses were shaped to solve problems that lasted not for years, but for seconds. They were primarily designed to get our muscles moving us as quickly as possible out of harm's way.

These days, our stresses are measured not in moments with mountain lions, but in hours, days, and sometimes months with hectic workplaces, screaming toddlers, and money problems. Our system isn't built for that. And when moderate amounts of stress hormones build up to large amounts, or hang around too long, they become quite harmful. That's how an exquisitely tuned system can become deregulated enough to affect a report card or a performance review—or a dog in a metal crate.

Cardiovascular system

Stress affects both our bodies and our brains, in both good and bad ways. Acute stress can boost cardiovascular performance— the probable source of those urban legends about grandmothers lifting one end of a car to rescue their grandchildren stuck under the wheels. Over the long term, however, too much adrenaline produces scarring on the insides of your blood vessels. These scars become magnets for molecules to accumulate, creating lumps called plaques. These can grow large enough to block the blood vessels. If it happens in the blood vessels of your heart, you get a heart attack; in your brain, you get a stroke. Not surprisingly, people who experience chronic stress have an elevated risk of heart attacks and strokes.

Immune system

Stress also affects our immune response. At first, the stress response helps equip your white blood cells, sending them off to fight on your body's most vulnerable fronts, such as the skin. Acute stress can even make you respond better to a flu shot. But chronic stress reverses these effects, decreasing your number of heroic white-blood-cell soldiers, stripping them of their weapons, even killing them outright. Over the long term, stress ravages parts of the immune system involved in producing antibodies. Together, these can cripple your ability to fight infection. Chronic stress also can coax your immune system to fire indiscriminately, even at targets that aren't shooting back—like your own body. Not surprisingly, people who experience chronic stress are sick more often. A *lot* more often. One study showed that stressed individuals were three times more likely to suffer from the common cold, especially if the stress was social in nature and lasted more than a month. They also are more likely to suffer from autoimmune disorders, such as asthma and diabetes.

To show how sensitive the immune system can be to stress, you need look no further than an experiment done at UCLA. Trained actors practiced Method acting, in which if a scene calls for you to be scared, you think of something frightening, then recite your lines while plumbing those memories. On one day, the actors performed using only happy memories. On another day, they performed using only sad memories. The researchers took blood samples, continually assessing their immune systems. On the "happy days," the actors had healthy immune systems. Their immune cells were plentiful, happy, readily available for work. On the "sad days," the actors showed something unexpected: a marked decrease in immune responsiveness. Their immune cells were not plentiful, not as robust, not as available to protect against infection.

Memory and problem solving

Stress affects memory. The hippocampus, that fortress of human memory, is studded with cortisol receptors like cloves in a ham. This makes it *very* responsive to stress signals. If the stress is not too severe, your brain performs better when it is stressed than when it is not stressed. You can solve problems more effectively and you are more likely to retain information. There's an evolutionary reason for this. Our survival on the savannah depended upon remembering what was life-threatening and what was not. Ancestors who could commit those experiences to memory the fastest (and recall them accurately with equal speed) were more apt to survive than those who couldn't. Indeed, research shows that memories of stressful experiences are formed almost instantaneously in the human brain, and they can be recalled very quickly during times of crises.

If the stress is too severe or too prolonged, however, stress begins to harm learning. Stressed people don't do math very well. They don't process language very efficiently. They have poorer memories, both short and long forms. Stressed people do not generalize or adapt old pieces of information to new scenarios as well as non-stressed individuals. They can't concentrate. In almost every way it can be tested, chronic stress hurts our ability to learn. One study showed that adults with high levels of stress performed 50 percent worse than adults with low levels of stress on tests of declarative memory (things you can declare) and executive function (the type of thinking that involves problem solving and self control). Those, of course, are the skills needed to excel in school, at work, and in relationships.

I remember a story by a flight instructor I knew well. He told me about the best student he ever had, and a powerful lesson he learned about what it meant to teach her. The student excelled in ground school. She aced the simulations, aced her courses. In the skies, she showed natural skill, improvising even in rapidly changing weather conditions. One day in the air, the instructor saw her doing something naïve. He was having a bad day and he yelled at her. He pushed her hands away from the airplane's equivalent of a steering wheel. He pointed angrily at an instrument. Dumbfounded, the student tried to correct herself, but in the stress of the moment, she made more errors, said she couldn't think, and then buried her head in her hands and started to cry. The teacher took control of the aircraft and landed it. For a long time, the student would not get back into the same cockpit. The incident hurt not only the teacher's professional relationship with the student but the student's ability to learn. It also crushed the instructor. If he had been able to predict how the student would react to his threatening behavior, he never would have acted that way. Relationships matter when attempting to teach human beings—whether you're a parent, teacher, boss, or peer. Here we are talking about the highly intellectual venture of flying an aircraft. But its success is fully dependent upon feelings.

The villain: cortisol

The biology behind this assault on our intelligences can be described as a tale of two molecules: one a villain, the other a hero. The villain is the aforementioned cortisol, part of a motley crew of stress hormones going by the name glucocorticoids. These hormones are secreted by the adrenal glands, which lie like a roof on top of your kidneys. The adrenal glands are so exquisitely responsive to neural signals, they appear to have once been a part of your brain that somehow fell off and landed in your mid-abdomen.

Stress hormones can do some truly nasty things to your brain if boatloads of the stuff are given free access to your central nervous system. And that's what is going on when you experience chronic stress. Stress hormones seem to have a particular liking for cells in the hippocampus, which is a problem because the hippocampus is deeply involved in many aspects of human learning. Stress hormones can make cells in the hippocampus more vulnerable to other stresses. Stress hormones can disconnect neural networks, the webbing of brain cells that store your most precious memories. For example, a bodyguard was in the car with Princess Diana on the night of her death. To this day, he cannot remember the events several hours before or after the car crash. Amnesia is a typical response to catastrophic stress. Its lighter cousin, forgetfulness, is quite common when the stress is less severe but more pervasive.

Stress hormones also can stop the hippocampus from giving birth to brand-new baby neurons. Under extreme conditions, stress hormones can even kill hippocampal cells. Quite literally, severe stress can cause brain damage in the very tissues most likely to help you succeed in life.

One of the most insidious effects of prolonged stress is that it pushes people into depression. I don't mean the "blues" people can experience as a normal part of daily living. Nor do I mean the grief resulting from tragic circumstance, such as the death of a relative. I am talking about the kind of depression that causes as many as 800,000 people a year to attempt suicide. It is a disease every bit as organic as diabetes, and often deadlier. Chronic exposure to stress can lead you to depression's doorstep, then push you through. Depression is a deregulation of thought processes, including memory, language, quantitative reasoning, fluid intelligence, and spatial perception. The list is long and familiar. But one of its hallmarks may not be as familiar, unless you are in depression. Many people who feel depressed also feel there is no way out of their depression. They feel that life's shocks are permanent and things will never get better. Even though there *is* a way out—treatment is often very successful—they have no perception of it. The situation feels so helpless that they don't seek treatment. Yet they can no more argue their way out of a depression than they could argue their way out of a heart attack. Clearly, stress hurts learning. Most important, however, stress hurts *people*.

The hero: BDNF

The brain seems to be aware of all this and has supplied our story not only with a villain but also with a hero. We met this champion in the Exercise chapter. It's brain-derived neurotrophic factor. BDNF is the premier member of a powerful group of proteins called neurotrophins. BDNF in the hippocampus acts like a peacekeeping force, keeping neurons alive and growing in the presence of hostile action. As long as there is enough BDNF around, stress hormones cannot do their damage.

How, then, does the system break down? The problem begins when too many stress hormones hang around in the brain too long, a situation you find in chronic stress, especially learned helplessness. As wonderful as the BDNF forces are, it is possible to overwhelm them if they are assaulted with a sufficiently strong (and sufficiently lengthy) glucocorticoid siege. Like a fortress overrun by invaders, enough stress hormones will eventually overwhelm the brain's natural defenses and wreak their havoc. In sufficient quantities, stress hormones are fully capable of turning off the gene that makes BDNF in hippocampal cells, causing long-lasting damage. You read that right: Not only can they overwhelm our natural defenses, but they can actually turn them *off*.

A genetic buffer

Out-of-control stress is bad news for the brains of most people. But of course "most" doesn't mean "all." Like oddly placed candles in a dark room, some people illuminate corners of human behavior with unexpected clarity. They illustrate the complexity of environmental and genetic factors.

Jill was born into an inner-city home. Her father began having sex with Jill and her sister during their preschool years. Her mother was institutionalized twice because of what used to be termed "nervous breakdowns." When Jill was 7 years old, her agitated dad called a family meeting in the living room. In front of the whole clan, he put a handgun to his head, said, "You drove me to this," and then blew his brains out. The mother's mental condition continued to deteriorate, and she revolved in and out of mental hospitals for years. When Mom was home, she would beat Jill. Beginning in her early teens, Jill was forced to work outside the home to help make ends meet. As Jill got older, we would have expected to see deep psychiatric scars, severe emotional damage, drugs, maybe even a pregnancy or two. Instead, Jill developed into a charming and quite popular young woman at school. She became a talented singer, an honor student, and president of her high-school class. By every measure, she was emotionally well-adjusted and seemingly unscathed by the awful circumstances of her childhood.

Her story, published in a leading psychiatric journal, illustrates the unevenness of the human response to stress. Psychiatrists long have observed that some people are more tolerant of stress than others. Molecular geneticists are beginning to shed light on the reasons. Some people's genetic complement naturally buffers them against the effects of stress, even the chronic type. Scientists have isolated some of these genes. In the future, we may be able to tell stress-tolerant from stress-sensitive individuals with a simple blood test, looking for the presence of these genes.

We each have our own tipping point

How can we explain the various ways humans respond to stress— both the typical cases and the exceptions? The answer is that stress is neutral. Aversive stimuli are neither beneficial nor bad. Whether stress becomes damaging depends on the severity of the stress, how long you are exposed to the stress, and on your body's ability to handle stress. There's a tipping point where stress becomes toxic. Scientist Bruce McEwen calls it the allostatic load. *Allo* is from a Greek word meaning variable; *stasis* means a condition of balance. McEwen's idea is that we have systems that keep us stable by constantly changing themselves. The stress system, with all of its intricacies, is one of those. The brain coordinates body-wide changes—from hormonal to behavioral changes—in response to the approach and retreat of potential threats.

Stress at home shows up at school

I know the allostatic load as the first time, and only time, I ever heard my mother swear. As you may recall, my mother was a fourth-grade teacher. I was upstairs in my room, unbeknownst to my mother, who was upstairs in her room grading papers. She was grading one of her favorite students, a sweet, brown-haired wisp of a girl I will call Kelly. Kelly was every teacher's dream kid: smart, socially poised, blessed with a wealth of friends. Kelly had done very well in the first half of the school year. The second half of the school year was another story, however. My mother sensed something was very wrong the moment Kelly walked into class after Christmas break. Her eyes were mostly downcast, and within a week she had gotten into her first fight. In another week, she got her first C on an exam, which would prove to be the high point, as her grades for the rest of the year fluttered between Ds and Fs. She was sent to the principal's office numerous times, and my mother, exasperated, decided to find out what caused this meltdown. She learned that Kelly's parents had decided to get a divorce over Christmas and that the family conflicts, from which the parents valiantly had insulated Kelly, had begun spilling out into the open. As things unraveled at home, things also unraveled at school. And on that snowy day, when my mother gave Kelly her third straight D in spelling, my mother also swore: "Damn it!" she said, nearly under her breath. I froze as she shouted, "THE ABILITY OF KELLY TO DO WELL IN MY CLASS HAS NOTHING TO DO WITH MY CLASS!"

She was describing the relationship between home life and school life, a link that has frustrated teachers for a long time. One of the greatest predictors of performance in school turns out to be the emotional stability of the home.

I have firsthand experience with the effects of stress on grades. I was a senior in high school when my mother was diagnosed with the disease that would eventually kill her. She had come home late from a doctor's visit and was attempting to fix the family dinner. But when I found her, she was just staring at the kitchen wall. She haltingly related the terminal nature of her medical condition and then, as if that weren't enough, unloaded another bombshell. My dad, who knew of Mom's condition, was not handling the news very well and had decided to file for divorce. I felt as if I had just been punched in the stomach. For a few seconds I could not move. School the next day, and for the next 13 weeks, was a disaster. I don't remember much of the lectures. I only remember staring at my textbooks, thinking that this amazing woman had taught me to read and love such books, that we used to have a happy family, and that all of this was coming to an end. What she must have been feeling, much worse than I could ever fathom, she never related. Not knowing how to react, my friends soon withdrew from me even as I withdrew from them. I lost the ability to concentrate, my mind continually wandering back to childhood. My academic effort became a train wreck. I got the only D I would ever get in my school career, and I couldn't have cared less.

Even after all these years, it is still tough to write about that time in my life. But it effectively illustrates Brain Rule #4: Stressed brains do not learn the same way as non-stressed brains.

My grief at least had an end point. In an emotionally unstable home, the stress seems never-ending. Consider the all-too-common case of children witnessing their parents fighting. The simple fact is that kids find unresolved marital conflict deeply disturbing. They cover their ears, stand motionless with clenched fists, cry, scowl, ask to leave, beg parents to stop. Study after study has shown that children—some as young as 6 months—react to adult arguments physiologically, such as with a faster heart rate and higher blood pressure. Kids of all ages who watch parents constantly fight have more stress hormones in their urine. They have more difficulty regulating their emotions, soothing themselves, and focusing their attention on others. They are powerless to stop the conflict, and the loss of control is emotionally crippling. As you know, perception of control is a powerful influence on the perception of stress. They are experiencing allostatic load.

Given that stress can powerfully affect learning, one might predict that children living in high-anxiety households would not perform as well academically as kids living in more nurturing households. That is exactly what studies show. Marital stress at home can negatively affect academic performance in almost every way measurable, and at nearly any age. Initial studies focused on grade-point averages over time, revealing striking disparities in achievement between kids whose parents are going through a divorce and control groups. Even when a couple stays together, children living in emotionally unstable homes get lower grades and do worse on standardized tests of math and reading. Careful subsequent investigations showed that it was the presence of overt conflict, not divorce, that predicted grade failure.

The stronger the degree of conflict, the greater the effect on performance. When teachers are asked to rate children's intelligence and aptitude, children from homes with conflict score lower. Such children are three times more likely to be expelled from school or to become pregnant as teenagers, and five times more likely to live in poverty. As social activist Barbara Whitehead put it, writing for the *Atlantic Monthly:* "Teachers find many children emotionally distracted, so upset and preoccupied by the explosive drama of their own family lives that they are unable to concentrate on such mundane matters as multiplication tables."

Physical health deteriorates; truancy and absenteeism increase. The absenteeism may occur because stress is depleting the immune system, which increases the risk of infection. Though the evidence is not as conclusive, a growing body of data suggests that children living in hostile environments are at greater risk for certain psychiatric disorders, such as depression and anxiety disorders. As children grow up, they can bring the effects of childhood stress into their own relationships and work lives.

Stress at work: too expensive to ignore

Lisa Nowak was a lethal combat pilot, decorated electronics warfare specialist, pretty, smart. The government spent millions of dollars training her to be an astronaut. She also was a mother with three kids on the verge of divorcing her husband one month before her biggest professional assignment: mission control specialist for a shuttle mission. Talk about built-up stress. She put some weapons in her automobile, grabbed a disguise, and even packed a bunch of adult diapers so that she didn't have to stop to use a bathroom. She then drove virtually nonstop from Houston to Orlando, allegedly to kidnap her target, a woman she thought was a threat to a fellow astronaut to whom she had taken a fancy. Instead of serving as the lead for one of America's most technically challenging jobs, this highly skilled engineer sat awaiting trial on attempted kidnapping and battery. Nowak later pled guilty to lesser charges and retired with a "less than honorable" discharge. She will never fly again, which makes this sad story nearly heartbreaking. It also makes the money spent on her training a colossal waste. But those few million dollars are minuscule compared with the cost of stress on the workplace as a whole.

The American Stress Institute estimates that American businesses lose $300 billion every year because of work-related stress. Sources of that loss include health-related costs, worker compensation bills, employee turnover, and absenteeism. That last item is a big deal. About one million people stay home from work every day because of stress (about 40% of all absences occur because of tension felt at work!). The Bureau of Labor Statistics found the average amount of time off due to stress was 20 days. That's costly. One day's absence costs the company about two times what the worker would make in that day. If the prolonged stress leads to depression, organizations are dealing with a direct assault on their intellectual capital. Depression hobbles fluid intelligence, problem-solving abilities (including quantitative reasoning), and memory formation. In a knowledge-based economy where intellectual dexterity is often *the* key to survival, that's bad news. Yet executives often give stress the shortest shrift.

What makes a workplace stressful

Three things matter in determining whether your workplace is stressful or productive: the type of stress you experience, the balance between stimulation and boredom in your job, and the condition of your home life.

The perfect storm of occupational stress appears to be a combination of two factors: (1) a great deal is expected of you, and (2) you have no control over whether you will perform well. This sounds like a formula for learned helplessness. On the positive side, restoration of control can return groups to productivity. Some companies are using a stress-reduction program involving increasingly popular mindfulness training. Mindfulness is a form of controlled meditation in which you learn to become aware of your environment without judging and learn to enjoy the moment, among other practices. A few companies tested the programs to see whether they work. They do. About 36 percent of the employees in an insurance company who enrolled in mindfulness training noticed a marked reduction in stress after taking the program. About 30 percent noticed an improvement in sleep. It has also been found to be effective against depression.

Control isn't the only factor in productivity. Employees on an assembly line, doing the same tired thing day after day, certainly can feel in control of their work processes. But the brain-numbing tedium can become a source of stress. What spices things up? Studies show that a certain amount of uncertainty can be good for productivity, especially for bright, motivated employees. What they need is a balance between controllability and uncontrollability. Slight feelings of uncertainty may cause them to deploy unique problem-solving strategies.

The third characteristic, if you are a manager, is none of your business. I am talking about workers' family lives. There's no such thing as a firewall between personal issues and work productivity. We don't have two brains that we can swap out depending upon whether we are in our office or in our living room. Stress in the workplace affects family life, causing more stress in the family. Stress in the family causes more stress at work, which in turn gets brought home again. It's a downward spiral, and researchers call it "work-family conflict." If you are a worker, you may have the most wonderful feelings about autonomy at work, and you may have tremendous problem-solving opportunities with your colleagues. But if your home life is a wreck, you can still suffer the negative effects of stress, and so can your employer.

Whether we look at school performance or job performance, we keep running into the profound influence of the emotional stability of the home. Is there anything we can do about something so fundamentally personal, given that its influence can be so terribly public? The answer, surprisingly, may be yes.

Marriage intervention

Famed marriage researcher John Gottman can predict the future of a relationship within three minutes of interacting with a couple. His ability to accurately forecast marital success or failure is close to 90 percent. His track record is confirmed by peer-reviewed publications. He may very well hold the future of the American education and business sectors in his hands.

How is he so successful? After years of careful observation, Gottman isolated specific marital behaviors—both positive and negative—that hold most of the predictive power. But this research was ultimately unsatisfying to a man like Gottman, akin to telling people they have a life-threatening illness but not being able to cure them. And so the next step in his research was to find a cure. Gottman devised a marriage intervention strategy based on improving the behaviors proven to predict marital success and eliminating the ones proven to predict failure. Even in its most modest forms, his intervention drops divorce rates by nearly 50 percent. What do his interventions actually do? They show couples how to decrease both the frequency and severity of their hostile interactions. This return to civility has many positive side effects besides marital reconstruction, especially if the couples have kids. And the couples often do have kids.

Gottman's marriage research invariably put him in touch with couples who were starting families. When the baby arrived, Gottman noticed that the couple's hostile interactions skyrocketed. Causes ranged from chronic sleep deprivation to the increased demands of a helpless new family member (little ones typically require that an adult satisfy some demand of theirs about three times a minute). By the time the baby was 1 year old, marital satisfaction had plummeted 70 percent. At the one-year mark, the risk for maternal depression had risen from 25 percent to a whopping 62 percent. The couples' risk for divorce increased, which meant American babies often were born into a turbulent emotional world.

That single observation gave Gottman and fellow researcher Alyson Shapiro an idea. What if he deployed his proven marital intervention strategies to married couples while the wife was pregnant? Before the hostility floodgates opened up? Before the depression rates went through the roof? Based on his years of research, he already knew the marriage would improve. The big question concerned the kids. What would an emotionally stable home environment do to the baby's developing nervous system? Gottman decided to find out.

The research investigation, deployed over several years, was called Bringing Baby Home. It consisted of exposing expectant couples to the marital interventions whether their marriages were in trouble or not, and then assessing the development of the child. Gottman and Shapiro uncovered a gold mine of information. They found that babies raised in the intervention households didn't look anything like the babies raised in the controls. Their nervous systems didn't develop the same way. Their behaviors weren't in the same emotional universe. Children in the intervention groups didn't cry as much. They had stronger attention-shifting behaviors. They responded to external stressors in remarkably stable ways. Physiologically, the intervention babies showed all the cardinal signs of healthy emotional regulation, while the controls showed all the signs of unhealthy, disorganized nervous systems. The differences were remarkable and revealed something hopeful and filled with common sense. By stabilizing the parents, Gottman and Shapiro were able to change not only the marriage but the child. I think Gottman's findings can change the world.

More ideas

What people do in their private life is their own business, of course. Unfortunately, what people do in their private life often affects the public. Consider the criminal history of a fellow who had recently moved from Texas to Washington. He absolutely *hated* his new home and decided to leave. Stealing the car of a neighbor (for the second time that month), he drove several miles to the airport and ditched the car. He then found a way to fool both the security officials and the gate managers and hopped a free ride back to Texas. He accomplished this feat a few months shy of his 10th birthday. Not surprisingly, this boy comes from a troubled home. And he is hardly alone. If something doesn't change the course of their lives, the private issue of raising such children soon will become a very public problem.

How can we capture this chapter's Brain Rule—stressed brains learn differently from non-stressed brains—and change the way we educate, parent, and do business? I have thought a lot about that.

Teach parents first

The current education system starts in first grade, typically around age 6. The curriculum is a little writing, a little reading, a little math. The teacher is often a complete stranger. And something important is missing. The stability of the home is completely ignored, even though it is one of the greatest predictors of future success at school. What if we took the home influence seriously?

My idea envisions an educational system where the first students are not the children but the parents. The curriculum? How to create a stable home life, using Gottman's powerful baby-nervous-system-changing protocols. The intervention could even start in a maternity ward, offered by a hospital (like a Lamaze class, which takes just about as much time). This would be a unique partnership between the health system and the education system. And it makes education, from the beginning of a child's life, a family affair.

A week after birth, parents and tots would engage in a curriculum designed around the amazing cognitive abilities of infants, from language acquisition to the powerful need for luxurious amounts of active playtime. Parents would learn things like how to talk with their babies and what types of objects help children learn about the physical world. (This is *not* a call to implement products in the strange industry that seeks to turn babies into Einsteins in the first year of life. Most of those products have not been tested, and some have been shown to be harmful to learning. My idea envisions a mature, rigorously tested pedagogy that does not yet exist—one more reason for educators and brain scientists to work together.) Along with this, parents would take an occasional series of marital refresher courses, just to ensure the stability of the home. Can you imagine what a child might look like academically after years of thriving in an emotionally stable environment? The child flourishes in this fantasy.

At the very least, couples (struggling or not) can seek out Gottman's research-based marriage intervention. They are readily available to individuals.

Free family counseling and child care

Historically, people have done their best work—sometimes world-changing work—in their first few years after joining the workforce. In the field of economics, most Nobel Prize–winning research is done in the first 10 years of the recipient's career. Albert Einstein published most of his creative ideas at the ripe old age of 26. It's no wonder that companies want to recruit young intellectual talent.

The problem in today's economy is that people typically are starting a family at the very time they are also supposed to be doing their best work. They are trying to be productive at some of the most stressful times of their lives. What if companies took this unhappy collision of life events seriously? They could offer Gottman's intervention as a benefit for every newly married, or newly pregnant, employee. It might reverse the negative flow of family stress that normally enters the workplace at this time in a person's life, enhance productivity, and perhaps even generate grateful, loyal employees.

Businesses also risk losing their best and brightest at this time, a decision especially hard on women. What if talented people didn't have to choose between career and family? Businesses could offer on-site child care and flexible work schedules simply to retain employees at the very time they are most likely to be valuable. As this affects women the most, businesses immediately would achieve more gender balance. My guess is that such an offering would so affect productivity that the costs of providing child care are offset by the gains. Not only might businesses create more stable employees in the current generation, they might be raising far healthier children for work in the next.

Power to the people

Plenty of books discuss how to manage stress, and the good ones all say the key is to get control back into your life. For individuals, that may mean leaving a stressful job or an abusive relationship.

Companies could detect work-related problems by developing a questionnaire based on Jeansok Kim and David Diamond's three-pronged definition of stress, to assess whether an employee feels powerless. The next step would be to change the situation.

It's no coincidence that stress researchers, education scientists, and business professionals come to similar conclusions about the effects of toxic stress on people. We have known most of the salient points since Marty Seligman stopped shocking those dogs in the mid-1970s. It is time we made productive use of that horrible line of research.

Exercise

Even if you're not experiencing the kind of out-of-control stress we've been discussing, you can minimize the stress in your daily life. Aerobic exercise, several times a week for 30 minutes each, is an excellent way to shore up your BDNF peacekeeping forces.

Brain Rule #4
Stressed brains
don't learn the same way.

• Your body's defense system—the release of adrenaline and cortisol—is built for an immediate response to a serious but passing danger, such as a saber-toothed tiger. Chronic stress, such as hostility at home, dangerously deregulates a system built only to deal with short-term responses.

• Under chronic stress, adrenaline creates scars in your blood vessels that can cause a heart attack or stroke, and cortisol damages the cells of the hippocampus, crippling your ability to learn and remember.

• Individually, the worst kind of stress is the feeling that you have no control over the problem—you are helpless.

• Emotional stress has huge impacts across society, on children's ability to learn in school and on employees' productivity at work.

wiring

Every brain is wired differently.

MICHAEL JORDAN'S ATHLETIC FAILURES are puzzling, don't you think? In 1994, one of the best basketball players in the world—ESPN's greatest athlete of the 20th century—decided to quit the game and take up baseball instead. It was an attempt to fulfill a childhood dream. Jordan failed miserably. He played only one full season, during which he posted a .202 batting average and committed 11 errors in the outfield: the league's worst. Jordan's performance was so poor, he couldn't even qualify for a triple-A farm team. Though it seems preposterous that anyone with his physical ability could fail at any athletic activity he put his mind to, here was proof that one could. That same year, another athletic legend, Ken Griffey Jr., was burning up the baseball diamond. Like Jordan, Griffey Jr. played in the outfield but, unlike Jordan, he was known for catches so spectacular he seemed to float in the air. Float in the *air*? Wasn't that the space Jordan was accustomed to inhabiting? But the sacred atmosphere of the baseball park refused to budge for Jordan, and he soon went back to what his brains and muscles did better than anyone else's, creating a legendary sequel to an already stunning basketball career. Griffey, then playing for the red-hot Seattle Mariners, went on to bat .300 for seven years in the 1990s and, in that same decade, slug out 382 home runs. He is still sixth on the all-time home-runs list.

What made the talents of these two athletes so specialized? What was going on with the way their brains communicated better with certain muscles than others? It has to do with how their brains were wired. To understand what that means, we will take a guided tour through the brain to watch what happens as it is learning. We will discuss the enormous role of one's experience in how one's brain develops—including the fact that identical twins having an identical experience will not emerge with identical brains. And we will discover that we each have a Jennifer Aniston neuron. I am not kidding.

Learning rewires your brain

When you learn something, the wiring in your brain changes. Eric Kandel is the scientist mostly responsible for showing that acquiring even simple pieces of information physically alters the structure of our neurons. Taken broadly, these physical changes result in the functional organization and reorganization of the brain. This is astonishing. The brain is constantly learning things, so the brain is constantly rewiring itself.

Kandel first discovered this fact not by looking at humans but by looking at sea slugs. He soon found, somewhat insultingly, that human nerves learn things in the same way slug nerves learn things. And so do lots of animals in between slugs and humans. Kandel shared a Nobel Prize in 2000 for his work in part because it described the thought processes of virtually every creature with the means to think.

What are these physical alterations? As neurons learn, they swell, sway, and split. They break connections in one spot, glide over to a nearby region, and form connections with their new neighbors.

Many others stay put, simply strengthening their electrical connections with each other, increasing the efficiency of information transfer. Indeed, at this very moment inside your brain, bits of neurons are moving around like reptiles: slithering to new spots, getting fat at one end or creating split ends. All so that you can remember a few things about Eric Kandel and sea slugs.

This line of scientific inquiry started long before Kandel. In the 18th century, the Italian scientist Vincenzo Malacarne did a surprisingly modern series of biological experiments. He trained a group of birds to do complex tricks, then killed them and dissected their brains. He found that his trained birds had more extensive folding patterns in specific regions of their brains than his untrained birds. Fifty years later, Charles Darwin noted similar differences between the brains of wild animals and their domestic counterparts. The brains of wild animals were 15 to 30 percent larger than those of their tame, domestic counterparts. It appeared that the cold, hard world forced the wild animals into a constant learning mode. Those experiences wired their brains much differently.

It is the same with humans. This can be observed in places ranging from New Orleans's Zydeco beer halls to the staid palaces of the New York Philharmonic—both the natural habitat of violin players. In violin players' brains, the neural regions that control their left hands, where complex, fine motor movement is required on the strings, look as if they've been gorging on a high-fat diet. These regions are enlarged, swollen, and crisscrossed with complex associations. By contrast, the areas controlling the right hand, which draws the bow, look positively anorexic, with much less complexity.

The brain acts like a muscle: The more activity you do, the larger and more complex it can become. Whether that equates to more intelligence is another issue, but one fact is indisputable: What you do in life physically changes what your brain looks like. You can wire and rewire your brain with the simple choice of which musical instrument—or professional sport—you play.

Where wiring starts: the humble cell

You have heard since grade school that living things are made of cells, and for the most part, that's true. There isn't much that complex biological creatures can do that doesn't involve cells. You may have little gratitude for cells' generous contribution to your existence, but the cells make up for your indifference by ensuring that you can't control them. For the most part, they purr and hum behind the scenes, content to supervise virtually everything you will ever experience, much of which lies outside your awareness. Some cells are so unassuming, they find their normal function only after they can't function. The surface of your skin, for example—all nine pounds of it—literally is deceased. This allows the rest of your cells to support your daily life free of wind, rain, and spilled nacho cheese at a baseball game. It is accurate to say that nearly every inch of your outer physical presentation to the world is dead.

Of the cells that are alive, most look just like fried eggs. The white of the egg we call the cytoplasm; the center yolk is the nucleus. The nucleus contains that master blueprint molecule, DNA. DNA possesses genes, small snippets of biological instructions, that guide everything from how tall you become to how you respond to stress. A lot of genetic material fits inside that yolk-like nucleus. Nearly six feet of the stuff are crammed into a space that is measured in microns. A micron is 1/25,000th of an inch, which means putting DNA into your nucleus is like taking 30 miles of ribbon and stuffing it into an eggshell.

One of the most unexpected findings of recent years is that DNA, or deoxyribonucleic acid, is not randomly jammed into the nucleus. Rather, DNA is folded into the nucleus in a complex and tightly regulated manner. The reason for this molecular origami: cellular career options. Fold the DNA one way and the cell will become a contributing member of your liver. Fold it another way and the cell will become part of your busy bloodstream. Fold it a third way and you get the all-important nerve cell—and the ability to read this sentence.

What does a nerve cell look like? Like an uprooted tree: a large mass of roots on one end, connected to a small mass of branches on the other. The root mass in a nerve cell is called the cell body, and within it lies the nucleus. The tips of the roots are called dendrites. The thin, connecting trunk is called an axon, and the smaller mass of branches is called the axon terminal.

Nerve cells—also called neurons—help to mediate something as sophisticated as human learning. To understand how, I would like to take you on a guided tour of a neuron, borrowing from a science-fiction movie I saw as a child. It was called *Fantastic Voyage*, written by Harry Kleiner and popularized afterward in a book by the legendary Isaac Asimov. In the movie, four people are shrunk to microscopic size, and they board a tiny submarine to explore the internal workings of the human body. We are going to do the same. We'll roam around inside a typical neuron and the watery world in which it is anchored. Let's steer over to the hippocampus, the structure in the center of the brain where short-term knowledge is converted to longer-term knowledge.

When our little ship enters the hippocampus, our eyes adjust to the darkness and we peer out the windows. It looks as if we've entered an ancient, underwater forest. Everywhere there are submerged jumbles of branches, limbs, and trunks. Suddenly we see flashes of light in the darkness: sparks of electric current run up and down the trunks. The forest is electrified! We are going to have to be careful. Occasionally, large clouds of chemicals erupt from one end of the tree trunks, after electricity has convulsed through them.

These are not trees. These are neurons, with some odd structural distinctions. Sliding alongside one of the trunks, for example, we realize that the "bark" seems surprisingly slick, like grease. That's because it *is* grease. In the balmy interior of the human body, the exterior of the neuron, the phospholipid bilayer, is the consistency of

Mazola oil. The neuron's interior structure is what gives it its shape, much as the human skeleton gives the body its shape. When we plunge into the interior of the cell, one of the first things we will see is this skeleton. So let's plunge.

It's instantly, insufferably overcrowded, even hostile, in here. Everywhere we have to navigate through a dangerous scaffolding of spiky, coral-like protein formations: the neural skeleton. Though these dense formations give the neuron its three-dimensional shape, many of the skeletal parts are in constant motion—which means we have to do a lot of dodging. Millions of molecules still slam against our ship, however, and every few seconds we are jolted by electrical discharges. We don't want to stay long.

We escape from one end of the neuron. Instead of perilously winding through sharp thickets of proteins, we now find ourselves free-floating in a calm, seemingly bottomless watery canyon. In the distance, we can see another neuron looming ahead. We are in the space between two neurons, called a synaptic cleft, and the first thing we notice is that we are not alone. We appear to be swimming with large schools of tiny molecules. They are streaming out of the neuron we just visited and thrashing helter-skelter toward the one we are facing. In a few seconds, they reverse themselves, swimming back to the neuron we just left. It instantly gobbles them up. These schools of molecules are called neurotransmitters, and they function like tiny couriers. Neurons use these molecules to communicate information across the synaptic cleft. The cell that releases them is called the presynaptic neuron, and the cell that receives them is called the postsynaptic neuron.

Neurons release these chemicals into the synapse usually in response to being electrically stimulated. The neuron that receives these chemicals then reacts negatively or positively. In something like a cellular temper tantrum, the neuron can turn itself off to the rest of the neuroelectric world—a process termed inhibition. Or the neuron can become electrically stimulated, allowing a signal to be transferred: "I got stimulated and I am passing on the good news to you." The neurotransmitters then return to the cell of origin, a process appropriately termed reuptake. When that cell gobbles them up, the system is reset and ready for another signal.

As we gaze at this underwater hippocampal forest, we notice several disturbing developments. Some of these branches appear to be swaying, snakelike. Occasionally, the end of one neuron swells up, greatly increasing in diameter. The terminal ends of other neurons split down the middle like forked tongues, creating two connection points where there was only one. Electricity crackles through these moving neurons at a blinding 250 miles per hour, some quite near us, with clouds of neurotransmitters filling the synaptic spaces as the electric current passes by.

What we should do now is take off our shoes and bow low in our submarine, for we are on Neural Holy Ground. We are observing the process of the human brain *learning*.

As we slowly spin our ship 360 degrees, we notice how complicated this forest is. Take the two neurons between which we are floating. We are between just two connection points, two dendrites. If you can imagine two trees being uprooted by giant hands, turned 90 degrees so that the roots face each other, and then moved close enough to almost touch, you can visualize the real world of two neurons interacting in the brain. And that's just the simplest case. Usually, thousands of neurons are jammed up against one another, all occupying a single small parcel of real estate in the brain. The branches form connections with one another in a nearly incomprehensible mass of confusion. Ten thousand points of connection is typical.

Frenetic growth and frantic pruning

How do we get so many neurons? Infants provide a front-row seat to one of the most remarkable construction projects on Earth. The human brain, only partially constructed at birth, won't be fully assembled for years. The biggest construction programs aren't finished until you are in your early 20s, with fine-tuning well into your 40s. When babies are born, their brains have about the same number of connections as adults have. That doesn't last long. By the time children are 3 years old, the connections in specific regions of their brains have doubled or even tripled. That doesn't last long, either. The brain soon takes thousands of tiny pruning shears and trims back a lot of this hard work. By the time children are 8 or so, they're back to their adult numbers. And if kids never went through puberty, that would be the end of the story. In fact, it is only the middle of the story. At puberty, the whole process begins again, but with different regions in the brain. Once again, you see frenetic neural outgrowth and furious pruning back. It isn't until parents begin thinking about college financial aid that children's brains begin to settle into their adult forms. From a connectivity point of view, there is a great deal of activity in the terrible twos and then, during the terrible teens, a great deal more.

Because this happens to every person at about the same time, it might seem like cellular soldiers are obeying growth commands in lockstep formation. But nothing approaching military precision is observed in the messy world of brain development. And it is at this imprecise point that brain development meets Brain Rule: Every brain is wired differently. Even a cursory inspection of the data reveals remarkable variation in growth patterns from one person to the next. Whether examining toddlers or teenagers, different regions in different children develop at different rates. There is a remarkable degree of diversity in the specific areas that grow and prune, and with what enthusiasm they do so.

I'm reminded of this whenever I see the class pictures from my wife's journey through the American school system. My wife went to school with virtually the same people for her entire K–12 experience (and actually remained friends with most of them). Comparing the kids to each other back then, I always shake my head in disbelief.

In the first-grade picture, the kids are all about the same age, but they don't look it. Some kids are short. Some are tall. Some look like mature little athletes. Some look as if they just got out of diapers. The girls almost always appear older than the boys. It's even worse in the junior-high pictures. Some of the boys look as if they haven't developed much since third grade. Others are clearly beginning to sprout whiskers. Some of the girls, flat-chested, look a lot like boys. Others look developed enough to make babies. And if we could look inside these kids' heads, we would see that their brains are *just as unevenly developed* as their bodies. Let's find out why.

The Jennifer Aniston neuron

Some of the neural connections you're born with have preset functions: they control basic housekeeping functions like breathing, heartbeat, your ability to know where your foot is even if you can't see it, and so on. Researchers call this "experience independent" wiring. The brain also holds off connecting neurons, waiting for external experience to direct it. "Experience expectant" wiring is related to areas such as visual acuity and perhaps language acquisition. And, finally, we have "experience dependent" wiring. It may best be explained by the following scene, which would be right at home in a grade B movie.

A man is lying on a surgical table, electrodes implanted in his brain to create a kind of GPS pinpointing electrical activity in the brain. The man needs to have some of his neural tissue removed—resected, in surgical parlance—because of life-threatening epilepsy, and the depth electrodes will help surgeons determine where the seizures are starting. The man is conscious. Suddenly, a researcher whips out a photo of Jennifer Aniston and shows it to the patient. A neuron in the man's head fires. The researcher lets out a war whoop.

This experiment really happened. The neuron in question responded to seven photographs of actress Jennifer Aniston, while it practically ignored the 80 other images of everything else, including famous and nonfamous people. Lead scientist Quian Quiroga said, "The first time we saw a neuron firing to seven different pictures of Jennifer Aniston—and nothing else—we literally jumped out of our chairs." There is a neuron lurking in your head that is stimulated only when Jennifer Aniston is in the room.

A Jennifer Aniston *neuron?* How could this be? Surely nothing in our evolutionary history suggests that Jennifer Aniston is a permanent denizen of our brain wiring. (Aniston wasn't even born until 1969, and there are regions in our brain whose designs are millions of years old). To make matters worse, the researchers also found a Halle Berry–specific neuron, a cell in a patient's brain that wouldn't respond to pictures of Aniston or anything else. Just Berry. A patient also had a neuron specific to Bill Clinton. It no doubt was helpful to have a sense of humor while doing this kind of brain research.

Welcome to the world of experience-dependent brain wiring, where a great deal of the brain is hardwired *not* to be hardwired. Like a beautiful, rigorously trained ballerina, we are hardwired to be flexible. We can immediately divide the world's brains into those who know of Jennifer Aniston or Halle Berry or Bill Clinton and those who don't. The brains of those who do are wired differently from those who don't. This seemingly ridiculous observation underlies a much larger concept. Our brains are so sensitive to external inputs that their physical wiring depends upon the culture in which they find themselves.

Even identical twins do not have identical brain wiring. Consider this thought experiment: Suppose two adult male twins rent the Halle Berry movie *Catwoman*, and we in our nifty little submarine are viewing their brains while they watch. Even though the twins are in the same room, sitting on the same couch, the twins see the movie from slightly different angles. We find that their brains are encoding visual memories of the video differently, in part because it is impossible to observe the video from the same spot. Seconds into the movie, they are already wiring themselves differently. One of the twins earlier in the day read a magazine story about panned action movies, a picture of Berry figuring prominently on the cover. While watching the video, this twin's brain is simultaneously accessing memories of the magazine story. We observe that his brain is busy comparing and contrasting comments from the text with the movie and is assessing whether he agrees with them. The other twin has not seen this magazine, so his brain isn't doing this. Even though the difference may seem subtle, the two brains are creating different memories of the same movie.

That's the power of the Brain Rule. Learning results in physical changes in the brain, and these changes are unique to each individual. Not even identical twins having identical experiences possess brains that wire themselves exactly the same way. Given this, can we know *anything* about the organ? Well, yes. The brain has billions of cells whose collective electrical efforts work in a similar fashion. Every human comes equipped with a hippocampus, a pituitary gland, and the most sophisticated thinking store of electrochemistry on the planet: a cortex. These tissues function the same way in every brain. How then can we explain the individuality? Consider a highway.

For each brain, a different road map

The United States has one of the most extensive and complex ground transportation systems in the world. There are lots of variations on the idea of "road," from interstate freeways, turnpikes, and state highways to residential streets, one-lane alleys, and dirt roads. Pathways in the human brain are similarly diverse. We have the neural equivalents of large interstate freeways, turnpikes, and state highways. These big trunks are the same from one person to the next, functioning in yours about the same way they function in mine. So a great deal of the structure and function of the brain is predictable. This may be the ultimate result of the double-humped growth and pruning program we talked of previously. That's the experience-independent wiring.

It's when you get to the smaller routes—the brain's equivalent of residential streets, one-laners and dirt roads—that individual patterns begin to show up. In no two people are they identical. That's the experience-dependent wiring. Every brain has a lot of these smaller paths, which is why the very small amounts to a big deal. It's why, for example, human intellect is so multifaceted. Psychologist Howard Gardner believes we have at least seven categories of intelligence: verbal/linguistic, musical/rhythmic, logical/mathematical, spatial, bodily/kinesthetic, interpersonal, and intrapersonal. It's a much broader idea of intelligence than the standard IQ test implies.

We can grasp the magnitude of each brain's differences by watching a skilled neurosurgeon at work. George Ojemann has a shock of white hair, piercing eyes, and the quiet authority of someone who for decades has watched people live and die in the operating room. He is one of the great neurosurgeons of our time, and he is an expert at a technique called electrical stimulation mapping.

Ojemann is hovering over the exposed brain of a man with severe epilepsy. The man's name is Neil. Ojemann is there to remove some of Neil's misbehaving brain cells. Before Ojemann takes anything out, however, he has to make a map. To do this, he needs to talk to Neil during surgery, so Neil is fully conscious. Fortunately, the brain has no pain receptors. Ojemann wields a thin silver wire, which sends out small, unobtrusive electrical shocks to anything it touches. If it brushed against your hand, you would feel only a slight tingling sensation. Ojemann gently touches one end of the wire to an area of his patient's brain. In the book *Conversations with Neil's Brain*, he describes what happens next:

> "Feel anything?"
> "Hey! Someone touched my hand," Neil volunteers. Neither the anesthesiologist nor I had come anywhere close to Neil's hand.
> "Which hand?" asks George.

"My right one, sort of like someone brushed the back side of it. It's still tingling a little." The right hand reports to the left side of the brain, and George evidently has located the hand area of [the] somatosensory cortex with the stimulator.

Ojemann marks the area by putting a small sterile piece of paper on it. He touches another spot. Neil says he feels something near his right cheek. Another tiny piece of paper. This call and response goes on for hours. Like a neural cartographer, Ojemann is mapping the various functions of his patient's brain, with special attention paid to the areas close to the epileptic tissue.

These are tests of the patient's motor skills. For reasons not well understood, however, epileptic tissues are often disturbingly adjacent to areas critical for language. So Ojemann also pays close attention to the regions involved in language processing, where words and sentences and grammatical concepts are stored. If the patient is bilingual, he will map critical language areas for both Spanish and English. He applies a paper dot marked *S* to the regions where Spanish exists, and he applies a small *E* where English is stored. Ojemann does this painstaking work with every single patient who undergoes this type of surgery. Why? The answer is a stunner. He has to map each individual's critical function areas because *he doesn't know where they are.*

Ojemann can't predict the function of very precise areas in advance of the surgery because no two brains are wired identically. Not in terms of structure. Not in terms of function. For example, from nouns to verbs to aspects of grammar, we each store language in different areas, recruiting different regions for different components. Bilingual people don't even store their Spanish and their English in similar places.

This individuality has fascinated Ojemann for years. He once combined the brain maps for 117 patients he had operated on over the years. Only in one region did he find a spot where most people had a critical language area, and "most" means 79 percent of the patients.

Data from electrical stimulation mapping give the most dramatic illustration of the brain's individuality. But Ojemann also wanted to know how stable these differences were during life, and if any of those differences predicted intellectual competence. He found interesting answers to both questions. First, the brain's road maps are established very early in life, and they remain stable throughout. Even if a decade or two had passed between surgeries, the brain region recruited to host a critical language area remained the same. Second, Ojemann found that structural differences were associated with performance on a language test (given before surgery). If patients performed poorly on the test, the wiring pattern of their critical language area tended to be widely distributed. It was tightly focused in patients who performed well on the test. Lower scores on the test also predicted that a patient's critical language area had taken up residence in the superior temporal gyrus, as opposed to another brain region. Again, experience had wired each brain differently, with real-world consequences.

More ideas

Does it make any sense that most schools expect every child to learn like every other? For example, we expect that kids should be able to read by age 6. Yet students of the same age show a great deal of intellectual variability. Studies show that about 10 percent of students do *not* have brains sufficiently wired to read at that age. And does it make any sense that most businesses strive to treat each employee the same, especially in a global economy replete with various cultural experiences? As you can guess, I don't think so. Here are a few ideas for aligning our schools and businesses with the way the brain works.

Smaller class size

All else being equal, it has been known for many years that smaller, more intimate schools create better learning environments than megaplex houses of learning. Smaller is better because a teacher can deeply understand the individual needs of only so many students. If you are a parent, you can look for (and lobby for) schools with smaller classes or a more favorable teacher-student ratio. A college student might consider attending a smaller school. A manager looking to train employees should do it in smaller groups.

Theory of Mind testing

As you may recall from the Introduction, Theory of Mind is about as close to mind reading as humans get. It is the ability to understand the interior motivations of someone else and the ability to construct a predictable "theory of how their mind works." Nearly all of us can do it, but some of us are better at it than others.

Theory of Mind skills give teachers critical knowledge about their students, a heightened sensitivity for when they are confused, when they are fully engaged, and when they have truly learned what is being taught. I have come to believe that people with advanced Theory of Mind skills possess the single most important ingredient for effectively communicating information. If I'm right, it's possible that the best teachers possess advanced Theory of Mind skills and the worst teachers don't.

In the future, Theory of Mind tests should be as standard as IQ tests. Schools and other organizations could use the tests to reveal the better teachers. Companies could include Theory of Mind tests as they screen for leaders. People considering careers as teachers or managers could take the tests to help them decide whether they're a good fit for the role.

Customized classrooms and workplaces

As an instructor teaches a class, students inevitably will experience learning gaps. Left untreated, these gaps cause students to fall further behind. Developers of educational apps are using software to determine where a student's competencies lie and then adaptively tailor exercises for the student in order to fill in any gaps. The effect is greatest when the software is integrated into a school program. In a large classroom, teacher alone or software alone is not as effective. I would like to see more research on this—as would parents and teachers anxious about the infiltration of tablets into classrooms. Studies should include typical and optimized student-teacher ratios.

Parents could embrace the apps and pay close attention to the effect on their kids. Parents could look for a school adopting the trend of a flipped classroom, where students review the lecture at home before class. Class time is instead spent on homework, and teachers give individualized help as needed. Parents who are financially able might choose schools organized around the idea that children learn different things at different speeds, such as Montessori schools. Students can supplement school classes with free online courses, which allow them to view and review material at their own pace, such as those available through Khan Academy.

As for employees working at organizations who treat all people the same way, it will be up to you to push for the things you value: the balance of vacation time versus pay, a flexible schedule, the way your role within the company works. If you're a manager, make a list of the cognitive strengths of your team. Some of your employees may be great at memorizing things. Others may be better at quantitative tasks. Some have good people skills. Some don't. Assigning work projects based on an employee's strengths may be critical to your group's productivity. You may discover you had a Michael Jordan on your team but couldn't see it because you were only asking him to play baseball.

Brain Rule #5
Every brain is wired differently.

• What you do and learn in life physically changes what your brain looks like—it literally rewires it.

• The various regions of the brain develop at different rates in different people.

• Neurons go through a growth spurt and pruning project during the terrible twos and teen years.

• No two people's brains store the same information in the same way in the same place.

• We have a great number of ways of being intelligent, many of which don't show up on IQ tests.

attention

Brain Rule #6
We don't pay attention
to boring things.

O IT WAS ABOUT THREE o'clock in the morning when I was startled into sudden consciousness by a small spotlight sweeping across the walls of our living room. In the moonlight, I could see the six-foot frame of a young man in a trench coat, clutching a flashlight and examining the contents of our house. His other hand held something metallic, glinting in the silvery light. As my sleepy brain was immediately and violently aroused, it struck me that my home was about to be robbed by someone younger than me, bigger than me, and in possession of a firearm. Heart pounding, knees shaking, I turned on the lights, went to stand guard outside my children's room, called the police, and prayed. Miraculously, a police car was in the vicinity and came within a minute of my phone call. This all happened so quickly that my would-be assailant left his get-away car in our driveway, engine still running. He was quickly apprehended.

That experience lasted only 45 seconds, but aspects of it are indelibly impressed in my memory, from the outline of the young man's coat to the shape of his firearm. My brain fully aroused, I will never forget the experience as long as I live.

The more attention the brain pays to a given stimulus, the more elaborately the information will be encoded—that is, learned—and retained. That has implications for employees, parents, and students. Whether you are an eager preschooler or a bored-out-of-your-mind undergrad, better attention always equals better learning. A multitude of studies, both old and new, show that paying attention improves retention of reading material, increases accuracy, and boosts clarity in writing, math, science, and every academic category that has ever been tested.

So I ask this question in every college course I teach: "Given a class of medium interest, not too boring and not too exciting, when do you start glancing at the clock, wondering when the class will be over?" There is always some nervous shuffling, a few smiles, then a lot of silence.

Eventually someone blurts out, "Ten minutes, Dr. Medina."

"Why 10 minutes?" I inquire.

"That's when I start to lose attention. That's when I begin to wonder when this torment will be over." The comments are always said in frustration. A college lecture is still about 50 minutes long.

Studies confirm my informal inquiry. Noted educator Wilbert McKeachie says in his book *Teaching Tips* that "typically, attention increases from the beginning of the lecture to 10 minutes into the lecture and decreases after that point." He's right. Before the first quarter hour is over in a typical presentation, people *usually* have checked out. If keeping someone's interest in a lecture were a business, it would have an 80 percent failure rate. What happens in the brain at the 10-minute mark to cause such trouble? Nobody knows. The brain seems to be making choices according to some stubborn timing pattern, undoubtedly influenced by both culture and gene. This fact suggests a teaching and business imperative: Find a way to get and hold somebody's attention for 10 minutes, then do it again.

But how? To answer that question, we will need to explore some complex pieces of neurological real estate. We are about to investigate the remarkable world of human attention—including what's going on in our brains when we turn our attention to something, the importance of emotions to attention, and the myth of multitasking.

Can I have your attention, please?

While you are reading this paragraph, millions of sensory neurons in your brain are firing simultaneously, all carrying messages, each attempting to grab your attention. Only a few will succeed in breaking through to your awareness, and the rest will be ignored either in part or in full. It is easy for you to alter this balance, effortlessly granting airplay to one of the many messages you were previously ignoring. (While still reading this sentence, can you feel where your elbows are right now?) The messages that do grab your attention are connected to memory, interest, and awareness.

Memory

What you pay attention to is often profoundly influenced by memory. In everyday life, you use your previous experiences to predict where you should pay attention.

Different environments create different expectations. This was profoundly illustrated by the scientist Jared Diamond in his book *Guns, Germs, and Steel.* He describes an adventure traipsing through the New Guinea jungle with native New Guineans. He relates that these natives tend to perform poorly at tasks Westerners have been trained to do since childhood. But they are hardly stupid. They can detect the most subtle changes in the jungle, good for following the trail of a predator or for finding their way back home. They know which insects to leave alone, know where food exists, and can erect and tear down shelters with ease. Diamond, who had never spent time in such places, has no ability to pay attention to these things. Were he to be tested on such tasks, he also would perform poorly.

Different cultures create different expectations as well. For example, *Science* magazine notes that "Asians pay more attention to context and to the relationships between focal (foreground) objects and background in their descriptions of visual scenes, whereas Americans mention the focal items with greater frequency." Such differences can affect how an audience perceives a given business presentation or class lecture.

Interest

If you have an interest in a subject or a person, or something is important to you, you tend to pay more attention to things related to that subject or person. That's why, if you get a certain breed of dog or buy a certain model of car, you suddenly start noticing the same dog or car everywhere you go. Your brain continuously scans the sensory horizon, constantly assessing events for their potential interest or importance. It gives the more important events extra attention.

Can the reverse occur, with attention creating interest? Marketing professionals think so. They have known for years that novel stimuli—the unusual, unpredictable, or distinctive—are powerful ways to harness attention in the service of creating interest. One example is a print ad for Sauza Conmemorativo tequila. It shows a single picture of an old, dirty, bearded man, donning a brimmed hat and smiling broadly, revealing a single tooth. Printed above the mouth is: "This man only has one cavity." A larger sentence below says: "Life is harsh. Your tequila shouldn't be." Flying in the face of most tequila marketing strategies, which consist of scantily clad 20-somethings dancing at a party, the ad is effective at using attention to create interest.

Awareness

Of course, we must be aware of something for it to grab our attention. A strange illustration of this comes from neurologist Oliver Sacks. He describes a wonderful older woman in his care: intelligent, articulate, and gifted with a sense of humor. She suffered a massive stroke in the back region of her brain that left her with a most unusual deficit: She could no longer pay attention to anything that was to her left. She could pick up objects only in the right half of her visual field. She could put lipstick only on the right half of her face. She ate only from the right half of her plate. This caused her to complain to the hospital nursing staff that her portions were too small! Only when the plate was turned and the food entered her right visual field could she pay any attention to it and have her fill.

How could this be? The brain can be divided roughly into two hemispheres of unequal function, and patients can get strokes in either. The hemispheres contain separate "spotlights" for visual attention. The left hemisphere's spotlight is small, capable of paying attention only to items on the right side of the visual field. The right hemisphere, however, has a global spotlight. According to Marsel Mesulam of Northwestern University, who made these discoveries, getting a stroke on your left side is much less catastrophic because your right side can pitch in under duress to aid vision.

Of course, sight is only one stimulus to which the brain is capable of paying attention. Just let a bad smell into the room for a moment, make a loud noise, touch someone's arm, or taste an unexpectedly bitter bite of food, and people easily will shift attention. We also pay close attention to our psychological interiors, mulling over internal events and feelings again and again with complete focus, with no obvious external sensory stimulation.

You can imagine how tough it is to research such an ephemeral concept. For one thing, we don't know the neural location of

consciousness, loosely defined as that part of the mind where awareness resides. The best data suggest that several systems are scattered
throughout the brain.

How the brain pays attention

What's going on in our heads when we turn our attention to something? Thirty years ago, a scientist by the name of Michael Posner
derived a theory that remains popular today. Posner started his
research career in physics, joining the Boeing Aircraft Company soon
out of college. His first major research contribution was to figure out
how to make jet-engine noise less annoying to passengers riding in
commercial airplanes. You can thank your relatively quiet airborne
ride, even if the screaming turbine is only a few feet from your ear-
drums, in part on Posner's first research efforts. His work on planes
eventually led him to wonder how the brain processes information
of any kind. This led him to a doctorate in research and to a powerful idea that's sometimes jokingly referred to as the Trinity Model.
Posner hypothesized that we pay attention to things using three sepa-
rable but fully integrated networks of neural circuitry in the brain.
I'll use a simple story to illustrate his model.

One pleasant Saturday morning, my wife and I were sitting on
our outdoor deck, watching a robin drink from our birdbath, when
all of a sudden we heard a loud "swoosh" above our heads. Looking
up, we caught the shadow of a red-tailed hawk, dropping like a thun-
derbolt from its perch in a nearby tree, grabbing the helpless robin
by the throat. As the raptor swooped by us, not three feet away, blood
from the robin splattered on our table. What started as a leisurely
repast ended as a violent reminder of the savagery of the real world.
We were stunned into silence.

In Posner's model, the brain's first system functions much like
the two-part job of a museum security officer: surveillance and alert.
He called it the Alerting or Arousal Network. It monitors the sensory environment for any unusual activities. This is the general level

of attention our brains are paying to our world, a condition termed "intrinsic alertness." My wife and I were using this network as we sipped our coffee, watching the robin. If the system detects something unusual, such as the hawk's swoosh, it can sound an alarm heard brain-wide. That's when intrinsic alertness transforms into specific attention, called phasic alertness.

After the alarm sounds, we orient ourselves to the attending stimulus, activating the second network: the Orienting Network. We may turn our heads toward the stimulus, perk up our ears, perhaps move toward (or away) from something. It's why both my wife and I immediately lifted our heads away from the robin, attending to the growing shadow of the hawk. The purpose is to gain more information about the stimulus, allowing the brain to decide what to do.

The third system, the Executive Network, controls what action we take next. Actions may include setting priorities, planning on the fly, controlling impulses, weighing the consequences of our actions, or shifting attention. For my wife and me, it was stunned silence, until one of us moved to clean off the blood.

So we have the ability to detect a new stimulus, the ability to turn toward it, and the ability to decide what to do based on its nature. Posner's model offered testable predictions about brain function and attention, leading to neurological discoveries that would fill volumes. Hundreds of behavioral characteristics have since been discovered as well. We'll focus on four that have considerable practical potential: emotions, meaning, multitasking, and timing.

Emotions get our attention

As the television advertisement opens, we see two men talking in a car. They are having a mildly heated discussion about one of them overusing the word "like" in conversation. As the argument continues, we notice out the passenger window another car barreling toward the men. It smashes into them. There are screams, sounds of shattering glass, quick-cut shots showing the men bouncing in the car, twisted metal. The final shot shows the men standing, in disbelief, outside their wrecked Volkswagen Passat. In a twist on a well-known expletive, these words flash on the screen: "Safe Happens." The spot ends with a picture of another Passat, this one intact and complete with its five-star side-crash safety rating. It is a memorable, even disturbing, 30-second spot.

That's because it's charged with emotion. Emotionally charged events are better remembered—for longer, and with more accuracy—than neutral events. While this idea may seem intuitively obvious, it's frustrating to demonstrate scientifically because the research community is still debating exactly what an emotion is. What we can say for sure is that when your brain detects an emotionally charged event, your amygdala (a part of your brain that helps create and maintain emotions) releases the chemical dopamine into your system. Dopamine greatly aids memory and information processing. You can think of it like a Post-it note that reads "Remember this!" Getting one's brain to put a chemical Post-it note on a given piece of information means that information is going to be more robustly processed. It is what every teacher, parent, and ad executive wants.

Certain events have an emotional charge only for specific people. For example, my brain pays a great deal of attention if someone is banging pots and pans. When my mother got angry (which was rare), she went to the kitchen, washing LOUDLY any dishes she discovered in the sink. And if there were pots and pans, she deliberately would crash them together as she put them away. This noise served to announce to the entire household (if not the city block) her displeasure at something. To this day, whenever I hear loudly clanging pots and pans, I experience an emotional stimulus—a fleeting sense of "You're in trouble now!" My wife, whose mother never displayed anger in this fashion, does not associate anything emotional with the noise of pots and pans. It's a John-specific stimulus.

But certain emotionally charged events are universal, capable of capturing the attention of all of us. Such stimuli come directly from our evolutionary heritage, so they hold the greatest potential for use in teaching and business. They are strictly related to survival concerns. Regardless of who you are, the brain pays a great deal of attention to several questions:

"Can I eat it? Will it eat me?"

"Can I mate with it? Will it mate with me?"

"Have I seen it before?"

Any of our ancestors who didn't remember threatening experiences thoroughly or acquire food adequately would not live long enough to pass on his or her genes. So the human brain has many dedicated systems exquisitely tuned to the perception of threat (that's why the robbery story grabbed your attention); to reproductive opportunity (sex sells); and to patterns (we constantly assess our environment for similarities, and we tend to remember things if we think we have seen them before).

One of the best TV spots ever made employed all three of those elements in an ever-increasing spiral. Steve Hayden produced the commercial, introducing the Apple computer in 1984. It won every major advertising award that year and set a standard for Super Bowl ads. The commercial opens onto a bluish auditorium filled with robot-like men all dressed alike. In a reference to the 1956 movie *1984*, the men are staring at a screen where a giant male face is spouting off platitude fragments such as "information purification!" and "unification of thought!" The men in the audience are absorbing these messages like zombies. Then the camera shifts to a young woman in gym clothes, sledgehammer in hand, running full tilt toward the auditorium. She is wearing red shorts, the only bright color in the entire commercial. Sprinting down the center aisle, she throws her sledgehammer at the screen containing Big Brother. The screen explodes in a hail of sparks and blinding light. Plain letters

flash on the screen: "On January 24th, Apple Computer will introduce Macintosh. And you'll see why 1984 won't be like *1984*."

All three elements are at work here. Nothing could be more
threatening to a country marinated in free speech than George
Orwell's *1984* totalitarian society. There is sex appeal, with the
revealing gym shorts, but there is a twist. Mac is a female, so-o-o
… IBM must be a male. In the female-empowering 1980s, a whopping statement on the battle of the sexes suddenly takes center stage.
Pattern matching abounds as well. Many people have read *1984* or
seen the movie. Moreover, people who were *really* into computers at
the time made the connection to IBM, a company often called Big
Blue for its suit-clad sales force. These universal emotional stimuli
are the reason why Apple's ad was so memorable.

Meaning before details

The brain pays more attention to the gist than to the peripheral
details of an emotionally charged experience. That's why, after
seeing Apple's *1984* ad, what you're most vividly left with is a general
impression of Apple. With the passage of time, our retrieval of gist
always trumps our recall of details. I am convinced that America's
love of retrieval game shows such as *Jeopardy!* exists because we are
dazzled by the unusual people who can invert this tendency.

Normally, if we don't know the gist—the *meaning*—of information, we are unlikely to pay attention to its details. The brain selects
meaning-laden information for further processing and leaves the rest
alone.

One simple way to harness this tendency is to present information in a logically organized, hierarchical structure. (Rain gear:
umbrella, raincoat, boots. Beach gear: sunglasses, swimsuit, sandals.) This allows people to derive the meaning of the words to one
another. Words presented this way are much better remembered
than words presented randomly (raincoat, sandals, sunglasses,
umbrella, swimsuit, boots)—typically 40 percent better.

John Bransford, a gifted education researcher, has spent many years studying what separates novice teachers from expert teachers. One of many things he noticed is the way the experts organize information. "[Experts'] knowledge is not simply a list of facts and formulas that are relevant to their domain; instead, their knowledge is organized around core concepts or 'big ideas' that guide their thinking about their domains," he cowrote in *How People Learn*.

If you want people to be able to pay attention, don't start with details. Start with the key ideas and, in a hierarchical fashion, form the details around these larger notions. Meaning *before* details.

The brain cannot multitask

Multitasking, when it comes to paying attention, is a myth. The brain naturally focuses on concepts sequentially, one at a time. At first that might sound confusing; at one level the brain does multitask. You can walk and talk at the same time. Your brain controls your heartbeat while you read a book. Pianists can play a piece with left hand and right hand simultaneously. Surely this is multitasking. But I am talking about the brain's ability to pay attention. It is the resource you forcibly deploy while trying to listen to a boring lecture at school. It is the activity that collapses as your brain wanders during a tedious presentation at work. This attentional ability is, to put it bluntly, not capable of multitasking.

As a professor, I've noticed a change in my students' abilities to pay attention to me during a lecture. They have a habit of breaking out their laptops while I'm talking. Three researchers at Stanford University noticed the same thing about the undergraduates they were teaching, and they decided to study it. First, they noticed that while all the students seemed to use digital devices incessantly, not all students did. True to stereotype, some kids were zombified, hyperdigital users. But some kids used their devices in a low-key fashion: not all the time, and not with two dozen windows open simultaneously. The researchers called the first category of students

Heavy Media Multitaskers. Their less frantic colleagues were called Light Media Multitaskers.

If you asked heavy users to concentrate on a problem while simultaneously giving them lots of distractions, the researchers wondered, how good was their ability to maintain focus? The hypothesis: Compared to light users, the heavy users would be faster and more accurate at switching from one task to another, because they were already so used to switching between browser windows and projects and media inputs. The hypothesis was wrong.

In every attentional test the researchers threw at these students, the heavy users did consistently worse than the light users. Sometimes dramatically worse. They weren't as good at filtering out irrelevant information. They couldn't organize their memories as well. And they did worse on every task-switching experiment. Psychologist Eyal Ophir, an author of the study, said of the heavy users: "They couldn't help thinking about the task they weren't doing. The high multitaskers are always drawing from all the information in front of them. They can't keep things separate in their minds." This is just the latest illustration of the fact that the brain cannot multitask. Even if you are a Stanford student in the heart of Silicon Valley.

To understand this conclusion, we must delve a little deeper into the third of Posner's trinity—the Executive Network. Let's look at what your Executive Network is doing as you, say, compose a long email and then get interrupted by a text message from your significant other.

Step 1: Shift alert

To write the email from a cold start, blood quickly rushes to your anterior prefrontal cortex. This area of the brain, part of the Executive Network, works just like a switchboard, alerting the brain that it's about to shift attention.

Step 2: Rule activation for task #1

The alert contains a two-part message, sent via electricity crackling throughout your brain. The first part is a search query to find the neurons capable of executing the writing task. The second part encodes a command that will rouse the neurons, once discovered. This process is called "rule activation," and it takes several tenths of a second to accomplish. You begin to write your email.

Step 3: Disengagement

While you're typing, the text message is picked up by your sensory systems—starting with your ears, if the phone dings, or your skin, if the phone vibrates in your pocket. Because the rules for writing a work email are different from the rules for texting a lover, your brain must disengage from the email-writing rules before you can respond. This occurs. The switchboard is consulted, alerting the brain that another shift in attention is about to happen.

Step 4: Rule activation for task #2

The brain deploys another two-part message seeking the rule-activation protocols for texting. As before, the first is a command to find the texting-lover rules, and the second is the activation command. Now you can message your significant other. As before, it takes several tenths of a second simply to perform the switch.

These four steps must occur in sequence *every time* you switch from one task to another. This takes time. *And it is sequential.* That's why we can't multitask. That's why people find themselves losing track of previous progress and needing to "start over," perhaps muttering things like "Now where was I?" each time they switch tasks. That's why a person who is interrupted takes 50 percent longer to accomplish a task and makes up to 50 percent more errors.

The best we can say is that people who appear to be good at multitasking actually have good working memories, capable of paying attention to several inputs *one at a time*. Some people, particularly younger people, are more adept at task switching. If a person is familiar with the tasks, the completion time and errors are much less than if the tasks are unfamiliar.

Still, taking your sequential brain into a multitasking environment can be like trying to put your right foot into your left shoe. A good example is driving while talking on a cell phone. Until researchers started measuring the effects of cell-phone distractions under controlled conditions, nobody had any idea how profoundly they can impair a driver. It's like driving drunk. Recall that large fractions of a second are consumed every time the brain switches tasks. Cell-phone talkers are more wild in their "following distance" behind the vehicle in front of them, a half second slower to hit the brakes in emergencies, and slower to return to normal speed after an emergency. In a half second, a driver going 70 mph travels 51 feet. Given that 80 percent of crashes happen within three seconds of some kind of driver distraction, increasing your amount of task switching increases your risk of an accident. More than 50 percent of the visual cues spotted by attentive drivers are missed by cell-phone talkers. Not surprisingly, they get in more wrecks than anyone except very drunk drivers. Putting on makeup, eating, and rubbernecking at an accident aren't much better. One study showed that simply reaching for an object while driving a car multiplies the risk of a crash or near-crash by nine times.

The brain needs a break

My parents hated the film *Mondo Cane* because of one disturbing scene: farmers force-feeding geese to make pâté de foie gras. Using fairly vigorous strokes with a pole, farmers literally stuffed food down the throats of these poor animals. When a goose wanted to regurgitate, a brass ring was fastened around its throat, trapping the food inside the digestive tract. Jammed over and over again, such nutrient oversupply eventually created a stuffed liver, pleasing to chefs around the world. Of course, it did nothing for the nourishment of the geese, who were sacrificed in the name of expediency.

My mother would often relate this story to me when she talked about what makes a good or bad teacher. "Most teachers overstuff their students," she would exclaim, "like those farmers in that awful movie!" When I went to college, I soon discovered what she meant. And now that I am a professor who has worked closely with the business community, I can see the habit close-up. The most common communication mistakes? Relating too much information, with not enough time devoted to connecting the dots. Lots of force-feeding, very little digestion. This does nothing for the nourishment of the listeners, whose learning is often sacrificed in the name of expediency.

At one level, this is understandable. Most experts are so familiar with their topic that they forget what it is like to be a novice. Even if they remember, experts can become bored with having to repeat the fundamentals over and over again. In college, I found that a lot of my professors, because they had to communicate at such elementary levels, were truly fed up with teaching. They seemed to forget that the information was brand-new to us, and that we needed the time to digest it, which meant a need for consistent breaks. How true indeed that expertise doesn't guarantee good teaching!

I have observed similar mistakes in sermons, boardrooms, sales pitches, media stories—anywhere information from an expert needs to be transferred to a novice.

More ideas

Do one thing at a time

The brain is a sequential processor, unable to pay attention to two things at the same time. Businesses and schools praise multitasking, but research clearly shows that it reduces productivity and increases mistakes. Try creating an interruption-free zone during the day—turn off your email, phone, and social-media sites—and see whether you get more done. If you have trouble untangling yourself, download software that blocks your access to certain websites for the amount of time that you specify.

Divide presentations into 10-minute segments

Remember my students who said they got bored only 10 minutes into a mediocre lecture? The 10-minute rule, which researchers have known for many years, provides a guide to creating presentations people can pay attention to. Here's the model I developed for giving a lecture, for which I was named the Hoechst Marion Roussel Teacher of the Year (awarded at one of the largest annual meetings in psychiatry).

I decided that every lecture I'd ever give would be organized in segments, and that each segment would last only 10 minutes. Each segment would cover a single core concept—always large, always general, *and always explainable in one minute*. The brain processes meaning before detail, and the brain likes hierarchy. Starting with general concepts naturally leads to explaining information in a hierarchical fashion. Give the general idea *first*, before diving into details, and you will see a 40 percent improvement in understanding.

Each class was 50 minutes, so I could easily burn through five large concepts in a single period. I would use the other nine minutes in the segment to provide a detailed description of that single general concept. The trick was to ensure that each detail could be easily traced back to the general concept with minimal intellectual effort. I would regularly pause to explicitly explain the link. This is like allowing the geese to rest between stuffings. In addition to walking through the lecture plan at the beginning of the class, I sprinkled liberal repetitions of "where we are" throughout the hour.

This prevents the audience from trying to multitask. If the instructor presents a concept without telling the audience where that concept fits into the rest of the presentation, the audience is forced to simultaneously listen to the instructor and attempt to divine where it fits into the rest of what the instructor is saying. This is the pedagogical equivalent of trying to drive while talking on a cell phone. Because it is impossible to pay attention to ANY two things at once, this will cause listeners a series of millisecond delays throughout the presentation.

Then came the hardest part. After 10 minutes had elapsed, I had to be finished with the core concept. Why did I construct my lecture that way? I knew that I initially had only about 600 seconds to earn the right to be heard—or the next hour would be useless. And I knew that I needed to do something after the 601st second to "buy" another 10 minutes.

Bait the hook

After 9 minutes and 59 seconds, the audience's attention is getting ready to plummet to near zero. If something isn't done quickly, the students will end up in successively losing bouts of an effort to stay with me. What do they need? Not more information of the same type. Not some completely irrelevant cue that breaks them from their train of thought, making the information stream seem disjointed, unorganized, and patronizing. They need something so compelling that they blast through the 10-minute barrier—something that triggers an orienting response toward the speaker and captures executive functions, allowing efficient learning.

Do we know anything so potentially compelling? We sure do. An emotionally charged stimuli. So, every 10 minutes in my lecture, I decided to give my audiences a break from the fire hose of information and send them a relevant emotional charge, which I now call "hooks." As I did more teaching, I found the most successful hooks always followed these three principles:

1) The hook has to trigger an emotion.

Fear, laughter, happiness, nostalgia, incredulity—the entire emotional palette can be stimulated, and all work well. I employ survival issues here, describing a threatening event, a reproductive event (tastefully), or something triggering pattern matching. Narratives can be especially strong, especially if they are crisp and to the point.

What exactly do these hooks look like? This is where teaching can truly become imaginative. Because I work with psychiatric issues, case histories explaining some unusual mental pathology often rivet students to the upcoming (and drier) material. Business-related anecdotes can be fun, especially when addressing lay audiences in the corporate world. I often illustrate a talk about how brain science relates to business by addressing its central problem: vocabulary. I like the anecdote of the Electrolux vacuum cleaner company, a privately held corporation in Sweden trying to break into the North American market. They had plenty of English speakers on staff, but no Americans. Their lead marketing slogan? "If it sucks, it must be an Electrolux."

2) The hook has to be relevant.

It can't be just any story or anecdote. If I simply cracked a joke or delivered some irrelevant anecdote every 10 minutes, the presentation seemed disjointed. Or worse: The listeners began to mistrust my motives; they seemed to feel as if I were trying to entertain them at the expense of providing information. Audiences are really good at detecting disorganization, and they can become furious if they feel patronized. Happily, I found that if I made the hook very relevant to the provided content, the group moved from feeling entertained to feeling engaged. They stayed in the flow of my material, even though they were really taking a break.

3) The hook has to go between segments.

I could place it at the end of the 10 minutes, looking backward, summarizing the material, repeating some aspect of content. Or I could place it at the beginning of the module, looking forward, introducing new material, anticipating some aspect of content. I found that starting a lecture with a forward-looking hook relevant to the entire day's material was a great way to corral the attention of the class.

When I started placing hooks in my lectures, I immediately noticed changes in the audience members' attitudes. First, they were still interested at the end of the first 10 minutes. Second, they seemed able to maintain their attention for another 10 minutes or so, as long as another hook was supplied at the end. I could win the battle for their attention in 10-minute increments.

But then, halfway through the lecture, after I'd deployed two or three hooks, I found I could skip the fourth and fifth ones and still keep their attention fully engaged. I have found this to be true for students in 1994, when I first used the model, and in my lectures to this day. Will my model work for you as well as it works for me? I can't guarantee it. All I know for sure is that the brain doesn't pay attention to boring things, and I am as sick of boring presentations as you are.

Brain Rule #6
We don't pay attention to boring things.

- The brain's attentional "spotlight" can focus on only one thing at a time: no multitasking.

- We are better at seeing patterns and abstracting the meaning of an event than we are at recording detail.

- Emotional arousal helps the brain learn.

- Audiences check out after 10 minutes, but you can keep grabbing them back by telling narratives or creating events rich in emotion.

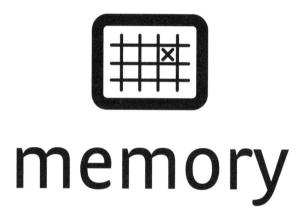

memory

Repeat to remember.

IT IS THE ULTIMATE intellectual flattery to be born with a mind so amazing that brain scientists voluntarily devote their careers to studying it. This impressive feat occurred with the owners of two such minds in the past century, and their remarkable brains provide much insight into human memory.

The first mind belongs to Kim Peek. He was born in 1951 with not one hint of his future intellectual greatness. He had an enlarged head, no corpus callosum, and a damaged cerebellum. He could not walk until age 4, and he could get catastrophically upset when he didn't understand something, which was often. Diagnosing him in childhood as mentally disabled, his doctors wanted to place him in a mental institution. That didn't happen, mostly because of the nurturing efforts of Peek's father, who recognized that his son also had some very special intellectual gifts. One of those gifts was memory. Peek had one of the most prodigious ever recorded. He could read two pages at the same time, one with each eye, comprehending and remembering perfectly everything contained in the pages. Forever.

Though publicity shy, Peek's dad once granted writer Barry Morrow an interview with his son. They met at a library, where Peek demonstrated to Morrow a familiarity with literally every book in the building. He then started quoting ridiculous—and highly accurate—amounts of sports trivia. After a long discussion about the histories of United States wars (Revolutionary to Vietnam), Morrow felt he had enough. He decided right then and there to write a screenplay about this man. Which he did: the Oscar-winning film *Rain Man*.

What was going on in the uneven brain of Kim Peek? Did his mind belong in a cognitive freak show, or was it only an extreme example of normal human learning? Clearly he had an extraordinary ability to remember facts. But something very important was occurring in the first few moments Peek's brain was exposed to information, and it's not so very different from what happens to the rest of us.

The first few moments of learning give us the ability to remember something. The brain has different types of memory systems, many operating in a semiautonomous fashion, and we know the most about declarative memory. Declarative memory involves something you can declare, such as "The sky is blue." It involves four steps: encoding, storing, retrieving, and forgetting. This chapter is about the first step. In fact, it is about the first few seconds of the first step. They are crucial in determining whether something that is initially perceived will also be remembered.

Why we have memory

We're not born knowing everything we need to know about the world. We must learn it through firsthand experience or secondhand teaching. Memory provides a big survival advantage. It allows us to remember where food grows and where threats lurk. For a creature as physically weak as humans (compare your fingernail with the claw of even a house cat, and weep with envy), not allowing experience to shape our brains would have meant almost certain death in the rough-and-tumble world of the savannah.

But memory is more than a Darwinian chess piece. Most researchers agree that its broad influence on our brains is what truly makes us consciously aware. The names and faces of our loved ones, our own personal tastes, and especially our awareness of those names and faces and tastes, are maintained through memory. We don't go to sleep and then, upon awakening, have to spend a week relearning the entire world. Memory does this for us. Even the single most distinctive talent of human cognition, the ability to write and speak in a language, exists because of active remembering. Memory, it seems, makes us not only durable but also human.

Types of memory

The type of memory Kim Peek was demonstrating so well is called declarative memory. You use it when you need to remember your Social Security number. Your retrieval commands might include things like visualizing the last time you saw the card, or remembering the last time you wrote down the number. And then you can state the number.

Here's how we know there's a second type of memory: Go ahead and remember how to ride a bike. Same process? Hardly. You do not call up a protocol list detailing where you put your foot, how to create the correct angle for your back, where your thumbs are supposed to be. The contrast proves an interesting point: One does not recall how to ride a bike in the same way one recalls nine numbers in a certain order. The ability to ride a bike seems quite independent from any conscious recollection of the skill. You were consciously aware when remembering your Social Security number, but not when remembering how to ride a bike. So declarative memories are those that can be experienced in our conscious awareness, such as a list of numbers, and nondeclarative memories are those that cannot be experienced in our conscious awareness, such as the motor skills necessary to ride a bike.

We also have both short-term forms of memory and long-term forms. A 19th-century German researcher was the first to show this. He performed the first real science-based inquiry into human memory—and he did the whole thing with his own brain. Hermann Ebbinghaus was born in 1850. As a young man, he looked like a cross between Santa Claus and John Lennon, with his bushy brown beard and round glasses. Ebbinghaus designed a series of experiments with which a toddler might feel at ease: He made up lists of nonsense words, 2,300 of them. Each word consisted of three letters and a consonant-vowel-consonant construction, such as TAZ, LEF, REN, ZUG. He then spent the rest of his life trying to memorize lists of these words in varying combinations and of varying lengths. With the tenacity of a Prussian infantryman (which, for a short time, he was), Ebbinghaus recorded his successes and failures. He uncovered many important things about human learning during this journey. He showed that memories have different life spans. Some memories hang around for only a few minutes, then vanish. Others persist for days or months, even for a lifetime. He uncovered one of the most depressing facts in all of education: People usually forget 90 percent of what they learn in a class within 30 days. And the majority of this forgetting occurs within the first few hours after class. Ebbinghaus also showed that one could increase the life span of a memory by repeating the information in timed intervals, something we'll talk about in the Memory chapter.

Long before we get to remembering or forgetting, there is a fleeting golden instant when the brain first encounters a new piece of declarative information. Let's see what the brain does.

We don't just press "record"

Tom was a blind teenager who could listen to complex pieces of music and then play them on the piano—on his first try—with the skill and artistry of a professional. He was so versatile on the instrument, he could simultaneously play a different song with each hand. Yet Tom never took piano lessons. In fact, Tom never took any kind of music lessons. He simply listened to other people play. When we hear about people like this, we are usually jealous. Tom absorbs music as if he could switch to the "on" position some neural recording device in his head. We think we also have this video recorder, only our model is not nearly as good. It is a common impression that the brain is a lot like a recording device: that learning is something akin to pushing the "record" button, and remembering is simply pushing "play." Wrong.

The initial moment of learning—of encoding—is incredibly mysterious and complex. The little we do know suggests that when information enters our head, our brain acts like a blender left running with the lid off. The information is chopped into discrete pieces and splattered all over the insides of our mind. This happens instantly. If you look at a complex picture, for example, your brain immediately extracts the diagonal lines from the vertical lines and stores them in separate areas. Same with color. If the picture is moving, the fact of its motion will be extracted and stored in a place separate than if the picture were static.

The brain slices and dices language the same way. One woman suffered a stroke in a specific region of her brain and lost the ability to use written vowels. You could ask her to write down a simple sentence, such as "Your dog chased the cat," and it would look like this:

$$Y \_ \_ r \, d \_ g \, ch \_ s \_ d \, t \, h \_ c \_ t.$$

There would be a place for every letter, but the vowels' spots were left blank! So we know that vowels and consonants are not stored in the same place. Her stroke damaged some kind of connecting wiring. Along the same lines, even though the woman lost the ability to fill in the vowels of a given word, she has perfectly preserved the place where the vowel should go. So the place where a vowel should go appears to be stored in a separate area from the vowel itself. Content is stored separately from its context/container. That is exactly the opposite of the strategy a video recorder uses to record things.

The blender

Why does this happen? To encode information means to convert data into, well, a code. Information is translated from one form into another so that it can be transmitted. From a physiological perspective, the brain must translate external sources of energy (sights, sounds, etc.) into electrical patterns the brain can understand. The brain then stores these patterns in separate areas. Here's an example.

One night I stayed with a friend who owned a beautiful lake cabin inhabited by a very large and hairy dog. Late next morning, I decided to go out and play fetch with this friendly animal. I made the mistake of throwing the stick into the lake and, not owning a dog in those days, had no idea what was about to happen to me. Like some friendly sea monster from Disney, the dog leapt from the water, ran at me full speed, suddenly stopped, then started to shake violently. With no real sense that I should have moved, I got sopping wet. To the brain, this story is all about energy and electricity.

My eyes picked up patterns of photons, or light, bouncing off the Labrador. Instantly, my brain converted them into patterns of electrical activity and routed the signals to the visual cortex in my occipital lobe. Now my brain can see the dog. In the initial moments of this learning, my brain transformed the energy of light into an electrical language it fully understands. My ears picked up the sound waves of the dog's loud bark. My brain converted the energy of the sound waves into the same brain-friendly electrical language. Then it routed them as well, but to the auditory cortex instead of the visual cortex. From a neuron's perspective, those two centers are a million miles away from each other. Any energy source—from the feel of the sun on my skin to the instant I unexpectedly and unhappily got soaked—goes through this conversion and routing process.

Encoding involves all of our senses, and their processing centers are scattered throughout the brain. Hence, the blender concept. In one 10-second encounter with an overly friendly dog, the brain recruits hundreds of different brain regions and coordinates the electrical activity of millions of neurons, encoding a single episode over vast neural differences.

Hard to believe, isn't it? The world appears to you as a unified whole. So how does your brain keep track of everything, and then how does it reunite all the elements to produce this perception of continuity? It is a question that has bothered researchers for years. It is called the "binding problem," from the idea that certain thoughts are bound together in the brain to provide continuity. We have very little insight into how the brain routinely and effortlessly gives us this illusion of stability.

Effortless vs. effortful processing

There's another way the brain decides how to encode information. Encoding when viewed from a psychological perspective is the manner in which we apprehend, pay attention to, and organize information so that we can store it. It is one of the many intellectual processes Kim Peek was so darn good at. The brain chooses among several types of encoding, and the ease which which we remember something depends in part on process used for encoding.

Automatic processing

Some years ago, I attended an amazing Paul McCartney concert. If you were to ask me what I had for dinner before the concert and what happened onstage, I could tell you about both events in great detail. Though the actual memory is very complex (composed of spatial locations, sequences of events, sights, smells, tastes, etc.), I did not have to write down some exhaustive list of its varied experiences, then try to remember the list in detail just in case you asked me about my evening.

This is because my brain deployed a type of encoding scientists call automatic processing. Automatic processing occurs with glorious unintentionality, requiring minimal attention or effort. The brain appears to use this type of encoding in cases where we can visualize the information we encounter. (Automatic processing is often associated with being able to recall the physical location of the information, what came before it, and what came after it.) It is very easy to recall data that have been encoded via this process. The memories seem bound all together into a cohesive, readily retrievable form.

Effortful processing

Automatic processing has an evil twin that isn't nearly so accommodating. As soon as the Paul McCartney tickets went on sale, I dashed to the purchasing website, which required my password for entrance. And I couldn't remember my password! Finally, I found the right one and snagged some good seats. But trying to commit these passwords to memory is quite a chore, and I have a dozen or so passwords written on countless lists, scattered throughout my house. Unlike my Social Security number, I don't use each password often enough to remember it. This kind of encoding—initiated deliberately, requiring conscious, energy-burning attention—is called effortful processing. The information does not seem bound together well at all, and it requires a lot of repetition before it can be retrieved with ease.

Others

Still other types of encoding exist. Three of them can be illustrated by taking the quick test below. Examine the capitalized word, and then answer the question below it.

FOOTBALL
Does this word fit into the sentence "I turned around to fight
_____"?

LEVEL
Does this word rhyme with evil?

MINIMUM
Are there any circles in these letters?

Answering each question requires very different intellectual skills, which researchers now know underlie different types of encoding. The first example illustrates semantic encoding: paying attention to the definitions of words. The second example illustrates phonemic encoding, involving a comparison between the sounds of words. The third example illustrates structural encoding. Simply asking for a visual inspection of shapes, it is the most superficial type.

You can see how the type of encoding your brain performs on a given piece of information would have a great deal to do with your ability to remember the information at a later date.

Cracking the code

All encoding processes share certain characteristics. If we heed two of them, we can better encode (and thus remember) information.

1) The more elaborately we encode information at the moment of learning, the stronger the memory.

When the initial encoding is more detailed, more multifaceted, and more embued with emotion, we form a more robust memory. You can demonstrate this right now with any two groups of friends. Have them gaze at the list of words below for a few minutes.

Tractor	Pastel	Airplane
Green	Quickly	Jump
Apple	Ocean	Laugh
Zero	Nicely	Tall
Weather	Countertop	

Tell Group #1 to determine the number of letters that have diagonal lines in them and the number that do not. Tell Group #2 to think about the meaning of each word and rate, on a scale of 1 to 10, how much they like or dislike the word. Take the list away, let a few minutes pass, and then ask each group to write down as many words as possible.

Which group remembers more words? The result you get has been replicated in laboratories many times over. As researchers Larry Squire and Eric Kandel write, "The result of the experiment is dramatic and consistent. The group that processed meaning remembers two to three times as many words as the group that focused on the shapes of the letters." You get the same result if you use pictures or even music.

At this point, you might be saying to yourself, "Well, duh!" Isn't it obvious that the more meaning something has, the more memorable it becomes? Most researchers would answer, "Well, yeah!" The very naturalness of the tendency proves the point. Hunting for diagonal lines in the word "apple" is not nearly as elaborate as remembering wonderful Aunt Mabel's apple pie, then rating the pie, and thus the word, a "10." The more personal, the better.

The trick for business professionals, and for educators, is to present information so compelling that the audience provides this meaning on their own, spontaneously engaging in deep and elaborate encoding.

2) The more closely we replicate the conditions at the moment of learning, the easier the remembering.

In one of the most unusual experiments performed in cognitive psychology, deep-sea divers were divided into two groups—one standing around on dry ground wearing wet suits and the other floating in about 10 feet of water, also wearing wet suits. Both groups of divers listened to somebody speak 40 random words. The divers then had to try to recall the list of words. The group that heard the words while in the water got a 15 percent better score if they were asked to recall the words while back in those same 10 feet of water, compared with standing on the beach. The group that heard the words on the beach got a 15 percent better score if they were asked to recall the words while suited on the beach, compared with floating in 10 feet of water.

Memory worked best, it appeared, if the environmental conditions at retrieval mimicked the environmental conditions at encoding. This occurs even under conditions where learning of any kind should be crippled, such as when a person is under the influence of marijuana and even laughing gas (nitrous oxide). Mood creates environmental conditions, too. Learn something while you are sad and you will be able to recall it better if, at retrieval, you are somehow suddenly made sad. It's called context-dependent or state-dependent learning. It may work because of the following concept.

One pathway for encoding and storing

After new information is perceived and processed, it is not transferred to some central hard drive in the brain for storage. There is no central hunting ground where memories go to be infinitely retrieved. Instead, the same neural pathways that the brain recruits to process new information are the same neural pathways that the brain uses to store the information. This means memories are distributed all over the surface of the cortex, with each brain region making its own contribution to a memory.

This idea is so counterintuitive that it may take an urban legend to explain it. At least, I think it's an urban legend. I heard it at a university administrators' luncheon I once attended. The keynote speaker told the story of the wiliest college president he ever encountered. The institute had completely redone its grounds in the summer, resplendent with fountains and beautifully manicured lawns. All that was needed was to install the sidewalks and walkways where the students could access the buildings. But there was no design for these permanent paths. The construction workers were anxious to install them and wanted to know what the design would be, but the president refused to give any. "Install them next year, please," he said. "I will give you the plans then." Disgruntled but compliant, the construction workers waited. The school year began, and the students were forced to walk on the grass to get to their classes. Very soon, defined trails started appearing all over campus, as well as large islands of beautiful green lawn. By the end of the year, the buildings were connected by paths in a surprisingly efficient manner. "Now," said the president to the contractors who had waited all year, "you can install the permanent sidewalks and pathways. Simply fill in all the paths you see before you!" The initial design, created by the initial input, also became the permanent path.

The brain's storage strategy is remarkably similar to the president's plan. New information penetrating the brain can be likened to the dirt paths that the students created across a pristine lawn. The final storage area can be likened to the pathways being permanently filled in with asphalt. They are the same pathways. This is why the initial moments of learning are so critical to retrieving that learning.

More ideas

The quality of the encoding stage—those earliest moments of learning—is one of the single greatest predictors of later learning success. We know that information is remembered best when it is elaborate, meaningful, and contextual. What can we do to take advantage of that in the real world?

First, we can take a lesson from a shoe store I used to visit as a little boy. This shoe store had a door with three handles at different heights: one near the very top, one near the very bottom, and one in the middle. The logic was simple: The more handles on the door, the more access points were available for entrance, regardless of the strength or age of customer. What a relief for a 5-year-old—a door I could actually reach! I was so intrigued with the door that I used to dream about it. In my dreams, however, there were hundreds of handles, all capable of opening the door to this shoe store.

"Quality of encoding" really means the number of door handles one can put on the entrance to a piece of information. The more handles one creates at the moment of learning, the more likely the information is to be accessed at a later date. The handles we can add revolve around content, timing, and environment.

Understand what the information means

The more a learner focuses on the meaning of information being presented, the more elaborately he or she will process the information. This principle is so obvious that it is easy to miss. What it means is this: When you are trying to drive a piece of information into your brain's memory systems, make sure you understand exactly what that information means. If you are trying to drive information into someone else's brain, make sure they understand exactly what it means. The corollary is true as well. If you don't know what the learning means, don't try to memorize the information by rote and pray the meaning will somehow reveal itself. And don't expect your students will do this either, especially if you have done an inadequate job of explaining things. This is like attempting to remember words by looking at the number of diagonal lines in the words.

Use real-world examples

How does one communicate meaning in such a fashion that learning is improved? A simple trick involves the liberal use of relevant real-world examples, thus peppering main learning points with meaningful experiences. As a student, you can do this while studying after class. Teachers can do it during the actual learning experience.

Numerous studies show this works. In one experiment, groups of students read a 32-paragraph paper about a fictitious foreign country.

The introductory paragraphs in the paper were highly structured. They contained either no examples, one example, or two or three consecutive examples of the main theme that followed. The greater the number of examples in the paragraph, the more likely the students were to remember the information. It's best to use real-world situations familiar to the learner. Remember wonderful Aunt Mabel's apple pie? That wasn't an abstract food cooked by a stranger; it was real food cooked by a loving relative. The more personal an example, the more richly it becomes encoded and the more readily it is remembered.

Examples work because they take advantage of the brain's natural predilection for pattern matching. Information is more readily processed if it can be immediately associated with information already present in the brain. We compare the two inputs, looking for similarities and differences as we encode the new information. Providing examples is the cognitive equivalent of adding more handles to the door. Providing examples makes the information more elaborative, more complex, better encoded, and therefore better learned.

Start with a compelling introduction

Introductions are everything. As an undergraduate, I had a professor who can thoughtfully be described as a lunatic. He taught a class on the history of cinema, and one day he decided to illustrate for us how art films traditionally depict emotional vulnerability. As he went through the lecture, he literally began taking off his clothes. He first took off his sweater and then, one button at a time, began removing his shirt, down to his T-shirt. He unzipped his trousers, and they fell around his feet, revealing, thank goodness, gym clothes. His eyes were shining as he exclaimed, "You will probably never forget now that some films use physical nudity to express emotional vulnerability. What could be more vulnerable than being naked?" We were thankful that he gave us no further details of his example. I will never forget the introduction to this unit in my film class (not that I'm endorsing its specifics). But its memorability illustrates the timing principle: The events that happen the first time you are exposed to information play a disproportionately greater role in your ability to accurately retrieve it at a later date. If you are trying to get information across to someone, a compelling introduction may be the most important single factor in the success of your mission. Why this emphasis on the initial moments? Because the memory of an event is stored in the same places initially recruited to perceive it.

Other professions have stumbled onto this notion. Budding directors are told by their film instructors that the audience needs to be hooked in the first three minutes after the opening credits to make the film compelling (and financially successful). Public speaking professionals say that you win or lose the battle to hold your audience in the first 30 seconds of a given presentation.

Create familiar settings

We know the importance of learning and retrieval taking place under the same conditions, but we don't have a solid definition of "same conditions." There are many ways for you to explore this idea.

One suggestion is that bilingual families create a "Spanish Room." This would be a room with a rule: Only the Spanish language could be spoken in it. The room could be filled with Hispanic artifacts and pictures of Spanish words. All Spanish would be taught there, and no English. Anecdotally, parents have told me this works.

When setting up their children's playroom at home, parents could create stations for science and stations for art—and not do science at the art station. Students could make sure that an oral examination is studied for orally, rather than by reviewing written material. Future car mechanics could be taught about engine repair in the actual shop where the repairs will occur.

At the moment of learning, environmental features—even ones irrelevant to the learning goals—may become encoded into the memory, right along with the goals. Environment then becomes part of elaborate encoding, the equivalent of putting more handles on the door.

After encoding, working memory kicks in

What happens to declarative information after those first few moments of encoding? We have the ability to hold it in our memory for a little while.

For many years, textbooks described this process using a metaphor involving cranky dockworkers, a large bookstore, and a small loading dock. An event to be processed into memory was likened to somebody dropping off a load of books onto the dock. If a dockworker hauled the load into the vast bookstore, it became stored for a lifetime. Because the loading dock was small, only a few loads could be processed at any one time. If someone dumped a new load of books on the dock before the previous ones were removed, the cranky workers simply pushed the old ones over the side.

Nobody uses this metaphor anymore. Short-term memory, we now know, is a much more active, much less sequential, far more complex process than that. Short-term memory is a collection of temporary memory capacities—busy work spaces where the brain processes newly acquired information. Each work space specializes in processing a specific type of information: auditory information, visual information, stories—plus a "central executive" to keep track of the activities of the others. These all operate in parallel. To reflect this multifaceted talent, short-term memory is now called working memory. The best way to explain working memory is to watch it in action. I can think of no better illustration than the professional chess world's first real rock star: Miguel Najdorf.

Rarely was a man more at ease with his greatness than Najdorf. He was a short, dapper fellow gifted with a truly enormous voice, and he had an annoying tendency to poll members of his audience on how they thought he was doing. Najdorf in 1939 traveled to a competition in Buenos Aires with the national team. Two weeks later, Germany invaded Najdorf's home country of Poland. Unable to return, Najdorf rode out the Holocaust tucked safely inside Argentina. He lost his parents, four brothers, and his wife to the concentration camps. Partly in hopes that any remaining family might read about it and contact him (and partly as a publicity stunt), he once played 45 games of chess simultaneously. He won 39 of these games, drew four, and lost two. While that is amazing in its own right, the truly phenomenal part is that he played all 45 games in all 11 hours blindfolded. You did not read that wrong. Najdorf never physically saw any of the chessboards or pieces; he played each game in his mind.

Several components of working memory were operating simultaneously in Najdorf's brain to allow him to do this. Najdorf's opponents verbally declared their chess moves. The work space assigned to linguistic information (called the phonological loop) allowed him to temporarily retain this auditory information.

To make his own chess move, Najdorf would visualize what each board looked like. The work space assigned to images and spatial input (called the visuospatial sketch pad) kicked in and allowed him to temporarily retain this visual information.

To separate one game from another, Najdorf's brain used the work space that keeps track of all activities throughout working memory (the central executive).

All of these work spaces have two things in common: All have a limited capacity, and all have a limited duration. Working memory is the bridge between the first few seconds of encoding and the process of storing a memory for a longer time. If the information held in working memory is not transformed into a more durable form, it will soon disappear.

What would happen if you lost the ability to convert short-term information to long-term memories? A 9-year-old boy, knocked off his bicycle, gave us an idea. Known to scientists as H.M., he is our second famous mind. The accident left H.M. with severe epilepsy. The seizures became so bad that, by his late 20s, H.M. was essentially a shut-in—a danger to himself and others. His family turned to famed neurosurgeon William Scoville in hopes of a cure. Scoville decided on drastic action: He would remove part of H.M's brain. The seizures were deemed to come from H.M.'s temporal lobe; if parts of it were removed, the logic went, the seizures should go away. The procedure, called a resection, is still in use today.

The surgeon won the battle but lost the war. The epilepsy was gone, but so was H.M.'s memory. He could meet you once and then meet you again an hour or two later, with absolutely no recall of the first visit. Even more dramatically, H.M. could no longer recognize his own face in the mirror. As his face aged, some of his physical features changed. But, unlike the rest of us, H.M. could not convert this new information into a longer-term form. This left him more or less permanently locked into a single idea about his appearance. When he looked in the mirror, he did not see this single idea, and he could not identify the person in the image. H.M.'s brain could still encode new information, but he had lost the ability to convert it.

The process of converting short-term memory traces to longer-term forms is called consolidation. It is our next subject.

Long-term memory

At first, a memory trace is flexible, labile, subject to amendment, and at great risk for extinction. Most of the inputs we encounter in a given day fall into this category. But some memories stick with us. Initially fragile, these memories strengthen with time and become remarkably persistent. They eventually reach a state where they appear to be infinitely retrievable and resistant to amendment. As we shall see, however, they're not as stable as we think. Nonetheless, we call these forms long-term memories. Consider the following story, which happened while I was watching a TV documentary with my then 6-year-old son. It was about dog shows. When the camera focused on a German shepherd with a black muzzle, an event that occurred when *I* was about his age came flooding back to my awareness.

In 1960, our backyard neighbor owned a dog he neglected to feed (we assumed) every Saturday. The dog bounded over our fence precisely at 8:00 a.m. every Saturday, ran toward our metal garbage cans, tipped out the contents, and began a morning repast. My dad got sick of this dog and decided one Friday night to electrify the can in such fashion that the dog would get shocked if his wet nose so much as brushed against it. Next morning, my dad awakened our entire family early to observe his "hot dog" show. To Dad's disappointment, the dog didn't jump over the fence until late in the morning, and he didn't come to eat. Instead, he came to mark his territory, which he did at several points around our backyard. As the dog moved closer to the can, my dad started to smile, and when the dog lifted his leg to mark our garbage can, my dad exclaimed, "Yes!" You don't have to know the concentration of electrolytes in mammalian urine to know that when the dog marked his territory on our garbage can, he also completed a mighty circuit. His cranial neurons ablaze, his reproductive future suddenly in serious question, the dog howled, bounding back to his owner. The dog never set foot in our backyard again; in fact, he never came within 100 yards of our house. Our neighbor's dog was a German shepherd with a distinct black muzzle, just like the one in the television show I was now watching. I had not thought of the incident in years.

What happened to my dog memory when summoned back to awareness? We used to think that consolidation, the mechanism that guides a short-term memory into a long-term memory, affected only newly acquired memories. Once the memory hardened, it never returned to its initial fragile condition. We don't think that anymore.

There is increasing evidence that when previously consolidated memories are recalled from long-term storage into consciousness, they revert to short-term memories. Acting as if newly minted into working memory, these memories may need to become reprocessed if they are to remain in a durable form.

That means my dog story is forced to start the consolidation process all over again, *every time I retrieve it*. This process is formally termed reconsolidation. As you can imagine, many scientists now question the entire notion of stability in human memory. If consolidation is not a sequential one-time event but an event that occurs every time a memory trace is reactivated, it means permanent storage exists in our brains only for those memories we choose not to recall! If this is true, the case I am about to make for repetition in learning is ridiculously important.

Retrieving memories: libraries and detectives

Like working memory, we appear to have different forms of long-term memory, most of which interact with one another. Unlike working memory, there is not as much agreement as to what those forms are. Most researchers believe we have semantic memory systems, in charge of remembering things like your sister's favorite dress or your weight in high school. Most believe there is episodic memory, in charge of remembering "episodes" of past experiences, complete with characters, plots, and time stamps—like your five-year high school reunion. Autobiographical memory, a subset of episodic memory, features a familiar protagonist: you.

How do we retrieve such memories? Two ways, researchers think. One model passively imagines libraries. The other aggressively imagines crime scenes.

In the library model, memories are stored in our heads the same way books are stored in a library. Retrieval begins with a command to browse through the stacks and select a specific volume. Once selected, the contents of the volume are brought into conscious awareness and read like a book. The memory is retrieved. This is the model we use soon after learning something (within minutes, hours, or days). In these cases, we are able to reproduce a fairly specific and detailed account of a given memory.

But as time goes by, and once-clear details fade, we switch to the second model. This model imagines our memories to be more like a large collection of crime scenes, complete with their own Sherlock Holmes. Retrieval begins by summoning the detective to a particular crime scene, full of fragments of data. Mr. Holmes examines the partial evidence available, and he invents a reconstruction of what was actually stored. The brain's Sherlock Holmes, however, isn't afraid to use a little imagination. In an attempt to fill in missing gaps, the brain relies on fragments, inferences, guesswork, and often— disturbingly—memories not even related to the actual event.

Why would the brain insert false information as it tries to reconstruct a memory? It stems from a desire to create organization out of a bewildering and confusing world. Here's what is happening: The brain constantly receives new inputs. It needs to store some of them in the same places already occupied by previous experiences. Trained in pattern matching, the brain connects new information to previously encountered information, in an attempt to make sense of the world. Accessing that previous information returns it to an amendable form. The new information resculpts the old. And the brain then sends the re-created whole back for new storage. What does this mean? Merely that present knowledge can bleed into past memories and become intertwined with them as if they were encountered together. Does that give you only an approximate view of reality? You bet it does.

Psychiatrist Daniel Offer demonstrated how faulty our Sherlock Holmes style of retrieval can be. If you had been one of his study subjects as a high-school freshman, Offer would have asked you to answer some questions that are really none of his business. Was religion helpful to you growing up? Did you receive physical punishment as discipline? Did your parents encourage you to be active in sports? And so on. Thirty-four years would go by. Offer then tracks you down and gives you the same questionnaire. Unbeknownst to you, he still has the answers you gave in high school, and he is out to compare your answers. How well do you do?

Horribly. Take the question about physical punishment, for example. Offer found that a third of the adults in his study recalled any physical punishment, such as spanking, as a kid. Yet nearly 90 percent of them had answered the question in the affirmative as adolescents.

Repetition fixes memories

Is there any hope of creating reliable long-term memories? As our Brain Rule—Repeat to remember—cheerily suggests, the answer is yes. Memory may not be fixed at the moment of learning, but repetition, doled out in specifically timed intervals, is the fixative.

Here's a test for you. Gaze at the following list of characters for about 30 seconds, then cover it up before you continue reading.

3 $ 8 ? A % 9

Can you recall the characters in the list without looking at them? Were you able to do it without internally rehearsing them? Don't be alarmed if you couldn't. The typical human brain can hold about seven pieces of new information for less than 30 seconds! If something does not happen in that short stretch of time, the information becomes lost. If you want to extend the 30 seconds to, say, a few minutes, or even an hour or two, you will need to consistently reexpose yourself to the information. This type of repetition is sometimes called maintenance rehearsal. It is good for keeping things in working memory—that is, for a short period of time. But there is a better way to push information into long-term memory. To describe it, I would like to relate the first time I ever saw somebody die.

Actually, I saw eight people die. The son of a career Air Force official, I was very used to seeing military airplanes in the sky. But I looked up one afternoon to see a cargo plane do something I had never seen before or since. It was falling from the sky, locked in a dead man's spiral. It hit the ground less than a thousand feet from where I stood, and I felt both the shock wave and the heat of the explosion. There are two things I could have done with this information. I could have kept it to myself, or I could have told the world. I chose the latter. After immediately rushing home to tell my parents, I called some of my friends. We met for sodas and began talking about what had just happened. The sounds of the engine cutting out. Our surprise. Our fear. As horrible as the accident was, we talked about it so much in the next week that the subject got tiresome. One of my teachers actually forbade us from bringing it up during class time, threatening to make T-shirts saying, "You've done enough talking."

Why do I still remember the details of this story? Because of my eagerness to yap about the experience. The gabfest after the accident forced a consistent reexposure to the basic facts, followed by a detailed elaboration of our impressions. This is called elaborative rehearsal, and it's the type of repetition most effective for the most robust retrieval. A great deal of research shows that thinking or talking about an event *immediately after it has occurred* enhances memory for that event, even when accounting for differences in type of memory. This is one of the reasons why it is so critical to have a witness recall information as soon as is humanely possible after a crime.

The timing of the repetitions is a key component. This was demonstrated by German researcher Hermann Ebbinghaus more than 100 years ago. He showed that repeated exposure to information *in spaced intervals* provides the most powerful way to fix memory into the brain.

Repetitions must be spaced out, not crammed in

Much like concrete, memory takes an almost ridiculous amount of time to settle into its permanent form. While it is hardening, it is maddeningly subject to amendment. As we discussed, new information can reshape or wear away previously existing memory traces. Such interference is likely to occur when we encounter an overdose of information without breaks, much like what happens in most conferences and classrooms. But this interference doesn't occur if the information is built up slowly, repeated in deliberately spaced cycles. Repetition cycles add information to our knowledge base, rather than disturbing the resident tenants.

If scientists want to know whether you are retrieving a vivid memory, they don't have to ask you. They can simply look in their fMRI machine and see whether your left inferior prefrontal cortex is active. Scientist Anthony Wagner used this fact to study two groups of students given a list of words to memorize. The first group was shown the words via mass repetition, reminiscent of students cramming for an exam. The second group was shown the words in spaced intervals over a longer period of time. The second group recalled the list of words with much more accuracy, with more activity in the cortex showing up on the fMRI (that's "functional magnetic resonance imaging) machine. Based on these results, Harvard psychology professor Dan Schacter wrote: "[I]f you want to study for a test you will be taking in a week's time, and are able to go through the material 10 times, it is better to space out the 10 repetitions during the week than to squeeze them all together."

Scientists aren't yet sure which time intervals supply all the magic. But taken together, the relationship between repetition and memory is clear. Deliberately re-expose yourself to information if you want to retrieve it later. Deliberately re-expose yourself to information *more elaborately* if you want to remember more of the details. Deliberately reexpose yourself to the information more elaborately and in fixed, spaced intervals if you want the retrieval to be as vivid as possible.

Memory consolidation goes fast, then slow

I was dating somebody else when I first met Kari—and so was she. But I did not forget Kari. She is a physically beautiful, talented, Emmy-nominated composer, and one of the nicest people I have ever met. When we both became "available" six months later, I immediately asked her out. We had a great time, and I began thinking about her more and more. Turns out she was feeling the same. Soon we were seeing each other regularly. After two months, it got so that every time we met, my heart would pound, my stomach would flip-flop, and I'd get sweaty palms. Eventually I didn't even have to see her to raise my pulse. Just a picture would do, or a whiff of her perfume, or … just *music*! Even a fleeting thought was enough to send me into hours of rapture. I knew I was falling in love.

What was happening to effect such change? With increased exposure to this wonderful woman, I became increasingly sensitive to her presence, needing increasingly smaller "input" cues (perfume, for heaven's sake?) to elicit increasingly stronger "output" responses. The effect has been long-lasting, with a tenure of more than three decades. Leaving the whys of the heart to poets and psychiatrists, the idea that increasingly limited exposures can result in increasingly stronger responses lies at the heart of how neurons learn things. Only it's not called romance; it's called long-term potentiation. LTP shows us how timed repetition works at the level of the neuron.

Fast consolidation

To describe LTP, we need to leave the world of behavior and drop down to the more intimate world of cell and molecule. Let's return to our tiny submarine in the hippocampus, where we were floating between two connected neurons. I will call the presynaptic neuron the "teacher" and the postsynaptic neuron the "student." The goal of the teacher neuron is to pass on information, electrical in nature, to the student cell. The teacher neuron, after receiving some stimulus, cracks off an electrical signal to its student. For a short period of time, the student becomes stimulated and fires excitedly in response. The synaptic interaction between the two is said to be temporarily "strengthened." This phenomenon is termed early LTP.

Unfortunately, the excitement lasts only for an hour or two. If the student neuron does not get the same information from the teacher within about 90 minutes, the student neuron's level of excitement will vanish. The cell will literally reset itself to zero and act as if nothing happened, ready for any other signal that might come its way. But if the information is repeatedly pulsed in discretely timed intervals—the timing for cells in a petri dish is three pulses, with about 10 minutes between each—the relationship between the teacher neuron and the student neuron begins to change. Much like my relationship with Kari after a few dates, increasingly smaller and smaller inputs from the teacher are required to elicit increasingly stronger and stronger outputs from the student. This response is termed late LTP.

When two neurons make it from early LTP to late LTP, you get synaptic consolidation. Scientists also call it fast consolidation, because it happens within minutes or hours. If it happens, that is. Any manipulation—behavioral, pharmacological, or genetic—that interferes with any part of this developing relationship will entirely block memory formation.

Slow consolidation

Two neurons alone don't allow us to form long-term memories. It's the fact that many neurons connect the hippocampus to the cortex, marrying the two in a chatty relationship. The cortex is that paper-thin layer of surface tissue that's about the size of a baby blanket when unfurled. The cortex is composed of six discrete layers of neural cells. These cells process signals originating from many parts of the body, including those lassoed by your sense organs. The cortex is connected to the deeper parts of the brain—including the hippocampus—by a hopelessly incomprehensible thicket of neural connections, like a complex root system. Communication between the cortex and hippocampus (lots of synaptic consolidation) is what allows the creation of long-term memories. This system consolidation takes a long time, so scientists call it slow consolidation.

Remember H.M., the man who couldn't recognize his own face in the mirror after his hippocampus was surgically removed? H.M. could meet you twice in two hours, with absolutely no recollection of the first meeting. He doesn't remember ever meeting a researcher who has worked with him for decades. This inability to encode information for long-term storage is called anterograde amnesia. H.M. also had retrograde amnesia, a loss of memory of the past. You could ask H.M. about an event that occurred three years before his surgery. No memory. Seven years before his surgery. No memory. If that's all you knew about H.M, you might conclude that his hippocampal loss created a complete memory meltdown. But you'd be wrong.

If you asked H.M. about the very distant past, say early childhood, he would display a perfectly normal recollection, just as you and I might. He can remember his family, where he lived, details of various events, and so on. This is a conversation with the researcher who studied him for many years:

Researcher: Can you remember any particular event that was special—like a holiday, Christmas, birthday, Easter?

H.M.: There I have an argument with myself about Christmastime.

Researcher: What about Christmas?

153

H.M.: Well, 'cause my daddy was from the South, and he didn't celebrate down there like they do up here—in the North. Like they don't have the trees or anything like that. And uh, but he came North even though he was born down Louisiana. And I know the name of the town he was born in.

H.M.'s childhood memory is intact starting about 11 years before his surgery. How is that possible? If the hippocampus is involved in all memory formation, removing the hippocampus should wipe the memory clean. But it doesn't. The hippocampus is relevant to memory formation for about 11 years after an event is recruited for long-term storage. After that, the memory somehow makes it to another region, one not affected by H.M.'s brain losses. Here's the interaction between the cortex and hippocampus that allows us to form long-term memories, and the reason H.M. still remembers Christmas:

1) The cortex receives sensory information and sends it to the hippocampus. They chat about it—a lot. Long after the initial stimulus has faded away, the hippocampus and the relevant cortical neurons are still yapping. As you sleep, the hippocampus is busy feeding signals back to the cortex, replaying a memory over and over again. The importance of sleep to learning is described in the Sleep chapter.

2) While the hippocampus and cortex are actively engaged, any memory they mediate is labile and subject to amendment.

3) After a period of time, the hippocampus will let go, effectively terminating the relationship with the cortex. The cortex is left holding the memory of the event. The hippocampus files for cellular separation only if the memory has become durable and fixed (consolidated) in the cortex. This process is at the heart of system consolidation, and it involves a complex reorganization of the brain regions supporting a particular memory trace.

How long does it take before the hippocampus lets go of its relationship with the cortex? In other words, how long does it take for a piece of information, once recruited for long-term storage, to become completely stable? Hours? Days? Months? The answer surprises nearly everybody who hears it for the first time. It can take *years*.

That's what the case of H.M., and patients like him, tell us. System consolidation, the process of transforming a short-term memory into a long-term one, can take years to complete. During that time, the memory is not stable.

As with short-term memories, long-term memories are stored in the same places that initially processed the stimulus. Retrieving a long-term memory 10 years later may simply be an attempt to reconstruct the initial moments of learning, when the memory was only a few milliseconds old!

Forgetting

We've talked about encoding, storage, and retrieval, the first three steps of declarative memory. The last step is forgetting. Forgetting plays a vital role in our ability to function for a deceptively simple reason. Forgetting allows us to prioritize. Anything irrelevant to our survival will take up wasteful cognitive space if we assign it the same priority as events critical to our survival. So we don't. At least, most of us don't.

Solomon Shereshevskii, a Russian journalist born in 1886, seemed to have a virtually unlimited memory. Scientists would give him a list of things to memorize, usually combinations of numbers and letters, and then test his recall. Shereshevskii needed only three or four seconds to "visualize" (his words) each item. Then he could repeat the lists back perfectly, forward or backward—even lists with more than 70 elements. In one experiment, developmental psychologist Alexander Luria exposed Shereshevskii to a complex formula of 30 letters and numbers. After a single recall test, which Shereshevskii accomplished flawlessly, the researcher put the list in a safe-deposit box *and waited 15 years*. Luria then took out the list, found Shereshevskii, and asked him to repeat the formula. Without hesitation, he reproduced the list on the spot, again without error.

Shereshevskii's memory of everything he encountered was so clear, so detailed, so *unending*, he lost the ability to organize it into meaningful patterns. Like living in a permanent snowstorm, he saw much of his life as blinding flakes of unrelated sensory information. He couldn't see the "big picture," meaning he couldn't focus on the ways two things might be related, look for commonalities, and discover larger patterns. Poems, carrying their typical heavy load of metaphor and simile, were incomprehensible to him. Shereshevskii couldn't forget, and it affected the way he functioned.

We have many types of forgetting, categories cleverly enumerated by researcher Dan Schacter in his book *The Seven Sins of Memory*. Tip-of-the-tongue lapses, absentmindedness, blocking habits, misattribution, biases, suggestibility—the list doesn't sound good. But they all have one thing in common. They allow us to drop pieces of information in favor of others. In so doing, forgetting helped us to conquer the Earth.

More ideas

Thinking and talking a lot about information soon after we encounter it (elaborate rehearsal) helps commit it to memory. Allowing time between repetitions is better than cramming. Unfortunately, we can't say exactly how much talking or exactly how much time produces the best result. You'll have to experiment.

I have some ideas about how we could systemically apply the concept of repetition in schools and companies.

Teaching in cycles

The day of a typical high-school student is segmented into five or six 50-minute periods, consisting of unrelenting, unrepeated streams of information. Here's my fantasy: In the school of the future, lessons are divided into 25-minute modules, cyclically repeated throughout the day. Subject A is taught for 25 minutes. Ninety minutes later, the 25-minute content of Subject A is repeated, and then a third time. *All* classes are segmented and interleaved in such a fashion.

Every third or fourth day would be reserved for quickly reviewing the facts delivered in the previous 72 to 96 hours. Students would inspect their notes, comparing them with what the teacher was saying in the review. This would result in a greater elaboration of the information and an opportunity to confirm facts. Because teachers wouldn't be able to address as much information, the school year would extend into the summer. Homework would be unnecessary, because students would already be repeating content during the day.

As I said, it's just a fantasy. Deliberately spaced repetitions have not been tested rigorously in the real world, so there are lots of questions. Do you really need three repetitions per subject per day to see a positive outcome? Do all subjects need such repetition? Would constant repetitions begin to interfere with one another as the day wore on? Do you even need the review sessions? We don't know.

Repetition over many years

Beyond doing well on the year-end test, our education system doesn't seem to care whether students actually remember what they learned. Given that system consolidation can take *years*, perhaps critical information should be repeated on a yearly or semiyearly basis.

In my fantasy class, this is exactly what happens. Take math. Repetitions begin with a review of multiplication tables, fractions, and decimals. Starting in the third grade, six-month and yearly review sessions occur through sixth grade. As students' competency grows, the review content becomes more sophisticated. But the cycles are still in place. I can imagine enormous benefits for every academic subject, especially foreign languages.

For businesses, I would extend the bachelor's degree into the workplace. You've probably heard that many corporations, especially in technical fields, are disappointed by the quality of the American undergraduates they hire. They have to spend money retraining many of their newest employees in basic skills that should have been covered in college.

I would turn your company into a learning and leadership factory, offering a full range of classes that would review every subject important to a new employee's job. Research would establish the optimal spacing of the repetition. More experienced employees might even begin attending these refresher courses, inadvertently rubbing shoulders with younger generations. The old guard would be surprised by how much they have forgotten, and how much the experience aids their own job performance.

I wish I could tell you this all would work. Instead, all I can say is that memory is not fixed at the moment of learning, and repetition provides the fixative.

Brain Rule #7
Repeat to remember.

→ The brain has many types of memory systems. Declarative memory follows four stages of processing: encoding, storing, retrieving, and forgetting.

→ Information coming into your brain is immediately fragmented and sent to different regions of the cortex.

→ The more elaborately we encode a memory during its initial moments, the stronger it will be.

→ You can improve your chances of remembering something if you reproduce the environment in which you first put it into your brain.

→ Working memory is a collection of busy work spaces that allows us to temporarily retain newly acquired information. If we don't repeat the information, it disappears.

→ Long-term memories are formed in a two-way conversation between the hippocampus and the cortex, until the hippocampus breaks the connection and the memory is fixed in the cortex—which can take years.

→ Our brains give us only an approximate view of reality, because they mix new knowledge with past memories and store them together as one.

→ The way to make long-term memory more reliable is to incorporate new information gradually and repeat it in timed intervals.

sensory
integration

Stimulate more of the senses.

EVERY TIME TIM SEES the letter *E*, he also sees the color red. He describes the color change as if suddenly forced to look at the world through red-tinted glasses. When Tim looks away from the letter *E*, his world returns to normal, until he encounters the letter *O*. Then the world turns blue. For Tim, reading a book is like living in a disco. For a long time, Tim thought this happened to everyone. When he discovered this happened to *no* one—at least no one he knew—he began to suspect he was crazy. Neither impression was correct. Tim is suffering—if that's the right word—from a brain condition called synesthesia. It's experienced by perhaps one in 2,000 people; some think more.

Synesthesia appears to be a short circuiting in the way the brain processes the world's many senses. But it also provides a strong clue that our sensory processes are wired to work together. In one of the strangest types of synesthesia—there are at least three dozen—people see a word and immediately experience a taste on their tongue. This isn't the typical mouthwatering response, such as imagining the taste of a candy bar after hearing the word "chocolate." This is like seeing the word "sky" in a novel and suddenly tasting a sour lemon in your mouth. A clever experiment showed that even when the synesthete could not recall the exact word, he or she would still get the taste from a general description of the word. Even when the brain's wiring gets confused, the senses still attempt to work together.

Here's another way we know the brain likes sensory integration. Suppose researchers show you a video of a person saying the surprisingly ugly syllable "ga." Unbeknownst to you, the scientists have turned off the sound of the original video and dubbed the sound "ba" onto it. When the scientist asks you to listen to the video with your eyes closed, you hear "ba" just fine. But if you open your eyes, your brain suddenly encounters the shape of the lips saying "ga" while your ears are still hearing "ba." The brain has no idea what to do with this contradiction. So it makes something up. If you are like most people, what you actually will hear when your eyes open is the syllable "da." This is the brain's compromise between what you hear and what you see—its need to attempt integration. It's called the McGurk effect.

But you don't have to be in a laboratory to see it in action. You can just go to a movie. The actors you see speaking to each other on-screen are not really speaking to each other at all. Their voices emanate from speakers cleverly placed around the room: some behind you, some beside you; none centered on the actors' mouths. Even so, you believe the voices are coming from those mouths. Your eyes observe lips moving in tandem with the words your ears are hearing, and the brain combines the experience to trick you into believing the dialogue comes from the screen. Together, these senses create the perception of someone speaking in front of you, when actually nobody is speaking in front of you.

The process of sensory integration has such a positive effect on learning that it forms the heart of Brain Rule #8: Stimulate more of the senses at the same time.

A fire hose of sights and sounds

An incredible amount of sensory information comes at us in any given moment. Imagine, for example, that you've gone out on a Friday night to a dance club in New York. The beat of the music dominates, hypnotic, felt more than heard. Laser lights shoot across the room. Bodies move. The smells of sweat, alcohol, and illegal smoking mix in the atmosphere like a second sound track. In the corner, a jilted lover is crying. You step out for a breath of fresh air. The jilted lover follows you. All of these external physical inputs and internal emotional inputs are presented to your brain in a never-ending fire hose of sensations. Does the example of a dance club seem extreme? It probably holds no more information than what you'd normally experience the next morning on the streets of Manhattan. Faithfully, your brain perceives the screech of the taxis, the smell of the pretzels for sale, the blink of the crosswalk signal, the touch of people brushing past. And your brain integrates them all into one coherent experience.

You are a wonder. We in brain-science land are only beginning to figure out how you do it.

It's mysterious: On one hand, your head crackles with the perceptions of the whole world—sight, sound, taste, smell, touch—as energetic as that dance party. On the other hand, the inside of your head is a darkened, silent place, lonely as a cave. The Greeks didn't think the brain did much of anything. They thought it just sat there like an inert pile of clay. Indeed, it does not generate enough electricity to prick your finger. Aristotle thought the heart held all the action. Pumping out rich, red blood 24 hours a day, the heart, he reasoned, harbored the "vital flame of life." This fire produced enough heat to give the brain a job description: to act as a cooling device. (He thought the lungs helped out, too.) Perhaps taking a cue from Aristotle, we still use the word "heart" to describe many aspects of mental life. Now we know that one of the brain's major job descriptions is to handle all of the inputs that our senses pick up and allow us to perceive the world.

How we perceive something

During the Revolutionary War, the British—steeped in the traditions of large European land wars—had lots of central planning. The field office gathered information from leaders on the battleground and then issued its commands. The Americans—steeped in the traditions of nothing—used guerrilla tactics: on-the-ground analysis and decision making prior to consultation with a central command. These very different approaches are a good way to describe the two main theories scientists have about how the brain goes from sensing something to perceiving it. Imagine the sound of a single gunshot over a green field during that war.

In the British model of this experience, our senses function separately, sending their information into the brain's central command, its sophisticated perception centers. Only in these centers does the brain combine the sensory inputs into a cohesive perception of the environment. The ears hear the rifle and generate a complete auditory report of what just occurred. The eyes see the smoke from the gun arising from the turf and process the information separately, generating a visual report of the event. The nose, smelling gunpowder, does the same thing. They each send their data to central command. There, the inputs are bound together, a cohesive perception is created, and the brain lets the soldier in on what he just experienced.

The American model puts things very differently. Here the senses work together from the very beginning, consulting and influencing one another quite early in the process. As the ears and eyes simultaneously pick up gunshot and smoke, the two impressions immediately confer with each other. They perceive that the events are occurring in tandem, without conferencing with any higher authority. The picture of a rifle firing over an open field emerges in the observer's brain. Perception is not where the integration begins but where the integration culminates.

Which model is correct? The data are edging in the direction of the second model. There are tantalizing suggestions that the senses do help one another, and in a precisely coordinated fashion. We'll talk about them in a couple of pages.

First sensing and routing, then perceiving

No matter which model is eventually declared the winner, the underlying processes are the same, and they operate in the same order: sensing, routing, and perceiving. Sensing involves capturing the energies from our environment that are pushing themselves into our orifices and rubbing against our skin. The brain converts this external information into a brain-friendly electrical language. Once the sensory information is encoded, it is routed to appropriate regions of the brain for further processing. As we discussed in the Wiring chapter, the signals for vision, hearing, touch, taste, and smell all have separate, specialized places where this processing occurs. A region called the thalamus—a well-connected, egg-shaped structure in the middle of your "second brain"—supervises most of this shuttling.

The information, dissected into sensory-size pieces and flung widely across the brain, next needs to be reassembled. Specialized areas throughout the brain take over from the thalamus to make this happen. They are not exactly sensory regions, and they are not exactly motor regions, but they are bridges between them. Hence, they are called association cortices. ("Cortices" is the plural of "cortex.") As sensory signals ascend through higher and higher orders of neural processing, the association cortices kick in.

The association cortices employ two types of processors: bottom up and top down. Let's walk through what they might be doing in your brain as you read the next sentence—a randomly chosen quote attributed to author W. Somerset Maugham.

The rank and file make a report

"There are only three rules for writing a novel," Maugham once said. "Unfortunately, nobody knows what they are."

After your eyes look at that sentence and your thalamus has routed each aspect of the sentence to the appropriate brain regions, "bottom-up" processors go to work. The visual system is a classic bottom-up processor. It has feature detectors that greet the sentence's visual stimuli. These detectors, working like auditors in an accounting firm, inspect every structural element in each letter of every word in Maugham's quote. They file a report, a visual conception of letters and words. An upside-down arch becomes the letter *U*. Two straight lines at right angles become the letter *T*. Combinations of straight lines and curves become the word "three." Written information has a lot of visual features in it, and this report takes a great deal of effort and time for the brain to organize. It is one of the reasons that reading is a relatively slow way to put information into the brain.

Higher-ups interpret the report

Next comes "top-down" processing. This can be likened to a board of directors reading the auditor's report and then reacting to it. Many comments are made. Sections are analyzed in light of pre-existing knowledge. The board in your brain has heard of the word "three" before, for example, and it has been familiar with the concept of rules since you were a toddler. Some board members have even heard of W. Somerset Maugham before, and they recall to your consciousness a movie called *Of Human Bondage*, which you saw in a film history course. Information is added to the data stream or subtracted from the data stream. In plenty of cases, as we saw in the McGurk effect, the brain resorts to making something up.

At this point, the brain generously lets you in on the fact that you are perceiving something.

Given that people have unique previous experiences, they bring different interpretations to their top-down analyses. Thus, two people can see the same input and come away with vastly different perceptions. It is a sobering thought. There is no one accurate way to perceive the world.

Smell is a powerful exception

Every sensory system must send a signal to the thalamus asking permission to connect to the higher levels of the brain where perception occurs—except for smell. Like an important head of state in a motorcade, nerves carrying information about smell bypass the thalamus and gain immediate access to their higher destinations.

Right between the eyes lies a patch of neurons about the size of a large postage stamp. This patch is called the olfactory region. The outer surface of this region, the one closest to the air in the nose, is the olfactory epithelium. When we sniff, odor molecules enter the nose chamber, penetrate a layer of snot, and collide with nerves there. The odor molecules brush against little quill-like protein receptors that stud the neurons in the olfactory epithelium. These neurons begin to fire excitedly, and you are well on your way to smelling something. The rest of the journey occurs in the brain.

One of the neurons' destinations is the amygdala. The amygdala supervises not only the formation of emotional experiences but also the *memory* of emotional experiences. Because smell directly stimulates the amygdala, smell directly stimulates emotions. Smell signals also beeline for a part of your brain deeply involved in decision making. It is almost as if the odor is saying, "My signal is so important, I am going to give you a memorable emotion. What are you going to do about it?"

Smell signals are in such a hurry, our receptor cells for smell aren't guarded by much of a protective barrier. This is different from most other sensory receptor cells in the human body. Visual receptor neurons in the retina are protected by the cornea, for example.

Receptor neurons that allow hearing in our ears are protected by the eardrum. The only things protecting receptor neurons for smell are boogers.

Pairing two senses boosts one

We've talked about the fact that the brain strives to integrate all of the senses, and we've touched on the regions of the brain involved in perceiving those senses. (We haven't talked about exactly *how* the brain integrates the senses, because, well, no one knows how that works.) Now let's look at those tantalizing hints that stimulating multiple senses at the same time increases the capability of the senses.

In one experiment, people watched a video of someone speaking, but with no sound. At the same time, scientists peered in on the brain using fMRI technology. The fMRI scans showed that the area of the brain responsible for processing the sound, the auditory cortex, was stimulated as if the person actually were hearing sound. If the subject was presented with a person simply "making faces," the auditory cortex was silent. It had to be a visual input *related* to sound. Then, visual inputs influence auditory inputs.

In another experiment, researchers showed short flashes of light near the subjects' hands, which were rigged with a tactile stimulator. Sometimes researchers would stimulate the subjects' hands while the light flashed, sometimes not. No matter how many times they did this, the visual portion of the brain always lighted up the strongest when the tactile response was paired with it. They could literally get a 30 percent boost in the visual system by introducing touch. This effect is called multimodal reinforcement.

Multiple senses affect our ability to detect stimuli, too. Most people, for example, have a very hard time seeing a flickering light if the intensity of the light is gradually decreased. Researchers decided to test that threshold by precisely coordinating a short burst of sound with the light flickering off. The presence of sound actually changed the threshold. The subjects found that they could see the light way beyond their normal threshold if sound was part of the experience.

Why does the brain have such powerful integrative instincts? The answer seems a bit obvious: The world is multisensory and has been for a very long time. Our East African crib did not unveil its sensory information one sense at a time during our development. Our environment did not possess *only* visual stimuli, like a silent movie, and then suddenly acquire an audio track a few million years later, and then, later, odors and textures. By the time we came down out of the trees, our evolutionary ancestors were already champions at experiencing a multisensory world. So it makes sense that in a multisensory environment, our muscles react more quickly, our eyes react to visual stimuli more quickly, and our threshold for detecting stimuli improves.

A multisensory environment enhances learning

Knowing that the brain cut its developmental teeth in an overwhelmingly multisensory environment, you might hypothesize that its learning abilities are increasingly optimized the more multisensory the situation is. You might further hypothesize that the opposite is true: Learning is less effective in a unisensory situation. That is exactly what you find.

Cognitive psychologist Richard Mayer probably has done more than anybody else to explore the link between multimedia exposure and learning. He sports a 10-megawatt smile, and his head looks exactly like an egg (albeit a very clever egg). His experiments are just as smooth: He divides the room into three groups. One group gets information delivered via one sense (say, hearing), another the same information from another sense (say, sight), and the third group the same information delivered as a combination of the first two senses.

The groups in the multisensory environments always do better than the groups in the unisensory environments. Their recall is more accurate, more detailed, and longer lasting—evident even

20 *years* later. Problem-solving ability improves, too. In one study, the group given multisensory presentations generated more than 50 percent more creative solutions on a problem-solving test than students who saw unisensory presentations. In another study, the improvement was more than 75 percent! Multisensory presentations are the way to go.

Many researchers think multisensory experiences work because they are more elaborate. Do you recall the counterintuitive concept that more elaborate information given at the moment of learning enhances learning? It's like saying that if you carry two heavy backpacks on a hike instead of one, you will accomplish your journey more quickly. But apparently our brains like heavy lifting. This is the "elaborative" processing that we saw in the Memory chapter. Stated formally, the extra cognitive processing of information helps the brain integrate the new material with prior information.

One more example of synesthesia supports this, too. Remember Solomon Shereshevskii's amazing mental abilities? He accurately reproduced a complex formula 15 years after seeing it once. Shereshevskii had multiple categories of (dis)ability. He felt that some colors were warm or cool, which is common. But he also thought the number one was a proud, well-built man, and that the number six was a man with a swollen foot—which is not common. Some of his imaging was nearly hallucinatory. He related: "One time I went to buy some ice cream … I walked over to the vendor and asked her what kind of ice cream she had. 'Fruit ice cream,' she said. But she answered in such a tone that a whole pile of coals, of black cinders, came bursting out of her mouth, and I couldn't bring myself to buy any ice cream after she had answered that way."

Synesthetes like Shereshevskii almost universally respond to the question "What good does this extra information do?" with an immediate and hearty, "It helps you remember." Most synesthetes report their odd experiences as highly pleasurable, which may, by virtue of dopamine, aid in memory formation. Indeed, synesthetes often have a photographic memory.

Smell boosts memory all by itself

I once heard a story about a man who washed out of medical school because of his nose. To fully understand his story, you have to know something about the smell of surgery, and you have to have killed somebody. Surgery can assault many of the senses. When you cut somebody's body, you invariably cut their blood vessels. To keep the blood from interfering with the operation, surgeons use a cauterizing tool, hot as a soldering iron. It's applied directly to the wound, burning it shut, filling the room with the acrid smell of smoldering flesh. Combat can smell the same way. And the medical student in question was a Vietnam vet with heavy combat experience. He didn't seem to suffer any aversive effects when he came home. He was accepted into medical school. But then the former soldier started his first surgery rotation. When he smelled the burning flesh from the cauterizer, it brought back to mind the immediate memory of an enemy combatant he had shot in the face, point-blank, an experience he had suppressed for years. The memory literally doubled him over. He resigned from the program the next week.

This story illustrates something scientists have known for years: Smell can evoke memory. It's called the Proust effect. Marcel Proust, the French author of the profoundly moving book *Remembrance of Things Past*, talked freely 100 years ago about smells and their ability to elicit long-lost memories. Why? Remember, smell neurons gain VIP access to the amygdala.

Smell has the unique advantage of being able to boost learning directly, without being paired with another sense. That's because it is an ancient sense, not fully integrated with the rest of the brain's sensory circuitry but instead closely wired to the emotional learning centers of the brain. In the typical experiment testing the effect of smell on remembering, two groups of people might be assigned to see a movie together and then told to report to the lab for a memory test. The control group takes the test in a plain room. The experimental group takes the test in a room smelling of popcorn. The second group blows away the first group in terms of number of events recalled, accuracy of events recalled, specific details, and so on. In some cases, they can accurately retrieve twice as many memories as the controls.

However, this is true for only certain types of memory. Odors appear to do their finest work when subjects are asked to retrieve the emotional details of a memory—as our medical student experienced—or to retrieve autobiographical memories. You get the best results if the smells are congruent. A movie test in which the smell of gasoline is pumped into the experimental room does not yield the same positive results as the smell of popcorn does.

Odors are not so good at retrieving declarative memory. Smell boosts declarative scores in only a couple of scenarios. It works if you're emotionally aroused—usually, that means mildly stressed—before the experiment begins. For some reason, showing a film of young Australian aboriginal males being circumcised is a favorite way to do this. And it works if you're asleep. Researchers used a version of a delightful card game my sons and I play on a regular basis. We use a deck of cards we purchased at a museum, resplendent with 26 pairs of animals. We turn all of the cards facedown, then start selecting two cards to find matches. It is a test of declarative memory. The one with the most correct pairs wins the game. In the experiment, the control groups played the game normally. But the experimental groups didn't. They played the game in the presence of rose scent. Then everybody went to bed. The control groups slept unperturbed. Soon after the snoring began in the experimental groups, however, the researchers filled their rooms with the same rose scent. Upon awakening, the subjects were tested on their knowledge of where the matches had been discovered the previous day.

The control group answered correctly 86 percent of the time. Those exposed and reexposed to the scent answered correctly 97 percent of the time. Brain imaging experiments showed the direct involvement of the hippocampus, that region of the brain deeply involved in memory. Smell, it appeared, enhanced recall during the offline brain processing that normally occurs during sleep.

Smell aside, there is no question that multiple cues, dished up via different senses, enhance learning. They speed up responses, increase accuracy, improve stimulation detection, and enrich encoding at the moment of learning.

More ideas

Multimedia presentations

Over the decades, Mayer has isolated a number of rules for multimedia presentations, linking what we know about working memory with his own empirical findings on how multimedia exposure affects human learning. Here are five of them, as he summarized in his book *Multimedia Learning*, useful for anyone giving a lecture, teaching a class, or creating a business presentation.

Multimedia principle: Students learn better from words and pictures than from words alone.

Temporal contiguity principle: Students learn better when corresponding words and pictures are presented simultaneously rather than successively.

Spatial contiguity principle: Students learn better when corresponding words and pictures are presented near to each other rather than far from each other on the page or screen.

Coherence principle: Students learn better when extraneous material is excluded rather than included.

Modality principle: Students learn better from animation and narration than from animation and on-screen text.

Sensory branding

Author Judith Viorst once said, "Strength is the capacity to break a [chocolate] bar into four pieces ... and then to eat just one of the pieces." No doubt, smell affects motivation. Can it also affect the motivation to buy?

One company tested the effects of smell on business and found a whopper of a result. Emitting the scent of chocolate from a vending machine, it found, drove chocolate sales up 60 percent. The same company installed a waffle-cone-smell emitter near a location-challenged ice cream shop: It was inside a large hotel and hard to find. Sales soared 50 percent, leading the inventor to coin the term "aroma billboard" to describe the technique.

Welcome to the world of sensory branding. Businesses are beginning to pay attention to human sensory responses, with smell as the centerpiece. For example, Starbucks does not allow employees to wear perfume on company time because it interferes with the seductive smell of the coffee they serve and its potential to attract customers.

Evidence for doing so comes from research by Dr. Eric Spangenberg, dean of the business school at Washington State University. Spangenberg knew from prior work that the male nose responds positively to the smell of rose maroc (spicy floral notes), the female nose to vanilla. What if he pumped rose maroc into the air of the men's section at a clothing store and vanilla into the women's section? Spangenberg hit pay dirt, generating twice the sales throughout the store. What if he then flipped the smells, introducing the male-preferred odor to the female section and vice versa?

Spangenberg hit pay dirt again: Sales went down. "You can't just use a pleasant scent and expect it to work," Spangenberg explained in an interview with *Fast Company*. "It has to be congruent."

Smell also can be used to differentiate a brand. Enter any Subway fast-food restaurant blindfolded and you'd instantly know where you were. In choosing a scent to represent your brand, one newspaper article advises, consider the aspirations of your potential buyer. Realtors sometimes employ the smell of freshly baked bread or cookies during an open house to remind buyers of the comforts of home, for example. Also match the odor to the "personality" of the object for sale, the article suggests. For potential buyers browsing an SUV dealership, the fresh scent of a forest or the salty odor of a beach might evoke a sense of adventure more so than, say, the scent of vanilla.

Research shows that the less complex the smell (the fewer interacting ingredients), the more likely it is to drive sales. Simpler smells drive sales 20 percent more than complex smells, or no smells at all.

Smells at work (not coming from the fridge)

I occasionally teach a molecular biology class for engineers, and I decided to do my own little Proust experiment. (There was nothing rigorous about this little parlor trick; it was simply an informal inquiry.) Every time I taught a section on the enzyme RNA polymerase II, I prepped the room by squirting the perfume Brut on one wall. In an identical class in another building, I taught the same material, but I did not squirt Brut when describing the enzyme. Then I tested everybody, squirting the perfume into both classrooms. Every time I did this experiment, I got the same result. The students who were exposed to the perfume during learning did better on subject matter pertaining to the enzyme—sometimes dramatically better—than those who were not. And that led me to an idea. Many businesses have a need to teach their clients about their products, from how to implement software to how to repair engines.

For financial reasons, the classes are often compressed in time and packed with information—90 percent of which is forgotten a *day* later. (For most declarative subjects, memory degradation starts the first few hours after the teaching is finished.) But what if you pair a smell with each lesson, as in my Brut experiment? Teachers could do this for the class as a whole, or you could do it on your own. You could spritz a bit of the scent near your pillow before you go to sleep, too. Overnight, you could not help but associate the autobiographical experience of the class—complete with the intense transfer of information—with the scent. Back at your company, when you need to apply what you learned, you could review your notes in the presence of the smell you encountered during the learning. See if it improves your performance, even cuts down on errors.

Is this context-dependent learning (remember those deep-sea divers from the Memory chapter) or a true multisensory environment? Either way, it's a start toward thinking about learning environments that go beyond our usual addiction to images and sounds.

Brain Rule #8

Stimulate more of the senses at the same time.

- We absorb information about an event through our senses, translate it into electrical signals (some for sight, others from sound, etc.), disperse those signals to separate parts of the brain, then reconstruct what happened, eventually perceiving the event as a whole.

- The brain seems to rely partly on past experience in deciding how to combine these signals, so two people can perceive the same event very differently.

- Our senses evolved to work together—vision influencing hearing, for example—which means that we learn best if we stimulate several senses at once.

- Smells have an unusual power to bring back memories, maybe because smell signals bypass the thalamus and head straight to their destinations, which include that supervisor of emotions known as the amygdala.

vision

Brain Rule #9
Vision trumps all other senses.

WE DO NOT SEE with our eyes. We see with our brains.

The evidence lies with a group of 54 wine aficionados. Stay with me here. To the untrained ear, the vocabularies that wine tasters use to describe wine may seem pretentious, more reminiscent of a psychologist describing a patient. ("Aggressive complexity, with just a subtle hint of shyness" is something I once heard at a wine-tasting soirée to which I was mistakenly invited—and from which, once picked off the floor rolling with laughter, I was hurriedly escorted out the door.)

These words are taken very seriously by the professionals, however. A specific vocabulary exists for white wines and a specific vocabulary for red wines, and the two are never supposed to cross. Given how individually we each perceive any sense, I have often wondered how objective these tasters actually could be. So, apparently, did a group of brain researchers in Europe. They descended upon ground zero of the wine-tasting world, the University of Bordeaux, and asked: "What if we dropped odorless, tasteless red dye into white wines, then gave it to 54 wine-tasting professionals?" With only visual sense altered, how would the enologists now describe their wine? Would their delicate palates see through the ruse, or would their noses be fooled? The answer is "their noses would be fooled." When the wine tasters encountered the altered whites, every one of them employed the vocabulary of the reds. The visual inputs overrode their other highly trained senses. Folks in the scientific community had a field day. Professional research papers were published with titles like "The Color of Odors" and "The Nose Smells What the Eye Sees." That's about as much frat-boy behavior as prestigious brain journals tolerate, and you can almost see the wicked gleam in the researchers' eyes. Studies such as these point to the nuts and bolts of Brain Rule #9. Visual processing doesn't just assist in the perception of our world. It dominates the perception of our world.

Not like a camera

Many people think that the brain's visual system works like a camera, simply collecting and processing the raw visual data provided by our outside world. Seeing seems effortless, 100 percent trustworthy, capable of providing a completely accurate representation of what's actually out there. Though we are used to thinking about our vision in such reliable terms, nothing in that last sentence is true. The process is extremely complex, seldom provides a completely accurate representation of our world, and is not 100 percent trustworthy. We actually experience our visual environment as a fully analyzed *opinion* about what the brain thinks is out there.

It starts with the retina, vying for the title of amateur filmmaker. We used to think the retina acted like a passive antenna in an automated process: First, light (groups of photons, actually) enters our eyes, where it is bent by the cornea, the fluid-filled structure upon which your contacts normally sit. The light travels through the eye to the lens, where it is focused and allowed to strike the retina, a group of neurons in the back of the eye. The collision generates electric signals in these cells, and the signals travel to the back of the brain via the optic nerve for analysis. But, it turns out, the retina isn't just waving through a series of unaltered electric signals. Instead, specialized nerve cells deep within the retina interpret the patterns of photons, assemble the patterns into a collection of "movies," and *then* send these movies for analysis. The retina, it seems, is filled with teams of tiny Martin Scorseses. These movies are called tracks.

Tracks are coherent, though partial, abstractions of specific features of the visual environment. One track appears to transmit a movie you might call *Eye Meets Wireframe*. It is composed only of outlines, or edges. Another makes a film you might call *Eye Meets Motion*, processing only the movement of an object (and often in a specific direction). Another makes *Eye Meets Shadows*. There may be as many as 12 of these tracks operating simultaneously in the retina, sending off interpretations of specific features of the visual field. This new view is quite unexpected. It's like discovering that the reason your TV gives you feature films is that your cable is infested by a dozen independent filmmakers, hard at work creating the feature while you watch it.

Rivers of visual information

These movies now stream out from the optic nerve, one from each eye, and flood the thalamus, that egg-shaped structure in the middle of our heads that serves as a central distribution center for most of our senses. If these streams of visual information can be likened to a large, flowing river, the thalamus can be likened to the beginning of a delta. Once the information leaves the thalamus, it travels along increasingly divided neural streams. Eventually, thousands of small neural tributaries will be carrying parts of the original information to the back of the brain. (Put your hand on the back of your head. Your palm is now less than a quarter of an inch away from the visual cortex, the area of the brain that is currently allowing you to see these words.) The information drains into a large complex region within the occipital lobe called the visual cortex.

Once they reach the visual cortex, the various streams flow into specific parcels. There are thousands of lots, and their functions are almost ridiculously specific. Some parcels respond only to diagonal lines, and only to specific diagonal lines (one region responds to a line tilted at 40 degrees, but not to one tilted at 45 degrees). Some process only the color information in a visual signal; others, only edges; others, only motion.

This means you can damage the region of the brain in charge of, say, motion, and get an extraordinary deficit. You'd be able to see and identify objects quite clearly, but not tell whether the objects are stationary or moving. This happened to a patient known to scientists as L.M. It's called cerebral akinetopsia, or motion blindness. L.M. perceives a moving object as a progressive series of still snapshots—like looking at an animator's drawings one page at a time. This can be quite hazardous. When L.M. crosses the street, for example, she can see a car, but she does not know if it is actually coming at her.

L.M.'s experience illustrates just how modular visual processing is. And if that was the end of the visual story, we might perceive our world with the unorganized fury of a Picasso painting—a nightmare of fragmented objects, untethered colors, and strange, unboundaried edges. But that's not what happens, because of what takes place next. The brain reassembles the scattered information. Individual tributaries start recombining, merging, pooling their information, comparing their findings, and then sending their analysis to higher brain centers. The centers gather these hopelessly intricate calculations from many sources and integrate them at an even more sophisticated level. Higher and higher they go, eventually collapsing into two giant streams of processed information. One of these, called the ventral stream, recognizes what an object is and what color it possesses. The other, termed the dorsal stream, recognizes the location of the object in the visual field and whether it is moving.

"Association cortices" do the work of integrating the signals. They associate—or, better to say, reassociate—the balkanized electrical signals. Then you see something. So the process of vision is not as simple as a camera taking a picture. The process is more complex and more convoluted than anyone could have imagined. There is no real scientific agreement about why this disassembly and reassembly strategy occurs.

Complex as visual processing is, things are about to get worse.

You're hallucinating right now

You might inquire whether I had too much to drink if I told you right now that you were actively hallucinating. But it's true. At this very moment, while reading this text, you are perceiving parts of this page that do not exist. Which means you, my friend, are hallucinating. I am about to show you that your brain actually likes to make things up, that it is not 100 percent faithful to what the eyes broadcast to it.

Blind spots

There is a region in the eye where retinal neurons, carrying visual information, gather together to begin their journey into deep brain tissue. That gathering place is called the optic disk. It's a strange region, because there are no cells that can perceive sight in the optic disk. It is blind in that region—and you are, too. It is called the blind spot, and each eye has one. Do you ever see two black holes in your field of view that won't go away? That's what you should see. But your brain plays a trick on you. As the signals are sent to your visual cortex, the brain detects the presence of the holes, examines the visual information 360 degrees around the spot, and calculates what is most likely to be there. Then, like a paint program on a computer, it fills in the spot. The process is called "filling in," but it could be called "faking it." Some scientists believe that the brain simply ignores the lack of visual information, rather than calculating what's missing. Either way, you're not getting a 100 percent accurate representation.

Dreams during the night—or day

It should not surprise you that the brain possesses an imaging system with a mind of its own. Proof is as close as your most recent dream. (There's a hallucination for you.) Actually, the visual system is even more of a loose cannon than that. Millions of people suffer from a phenomenon known as the Charles Bonnet Syndrome. Most who have it keep their mouth shut, however, and perhaps with good reason. People with Charles Bonnet Syndrome see things that aren't there.

Everyday household objects suddenly pop into view. Or unfamiliar people unexpectedly appear next to them at dinner. Neurologist Vilayanur Ramachandran describes the case of a woman who suddenly—and delightfully—observed two tiny policemen scurrying across the floor, guiding an even smaller criminal to a matchbox-size van. Other patients have reported angels, goats in overcoats, clowns, Roman chariots, and elves. The illusions often occur in the evening and are usually quite benign. Charles Bonnet Syndrome is common among the elderly, especially among those who previously suffered damage somewhere along their visual pathway. Interestingly, almost all of the patients know that the hallucinations aren't real.

A camel in each eye

Besides filling in our blind spots and creating bizarre dreams, the brain has another way of participating in our visual experience. We have two eyes, each taking in a full scene, yet the brain creates a single visual perception. Since ancient times, people have wondered why. If there is a camel in your left eye and a camel in your right eye, why don't you perceive two camels? Here's an experiment to try that illustrates the issue nicely.

1) Point your left index finger to the sky. Touch your nose and then stretch your left arm out.

2) Point your right index finger to the sky. Touch your nose and then move your finger about six inches away from your face.

3) Both fingers should be in line with each other, directly in front of your nose.

4) Now speedily wink your left eye and then your right one. Do this several times, back and forth. Your right finger will jump to the other side of your left finger and back again. When you open both eyes, the jumping will stop.

This little experiment shows that the two images appearing on each retina always differ. It also shows that both eyes working together give the brain enough information to create one stable picture. One camel. Two non-jumping fingers. How?

The brain interpolates the information coming from both eyes. Just to make things more complicated, each eye has its own visual field, and they project their images upside down and backward. The brain makes about a gazillion calculations, then provides you its best guess. And it is a guess. You can actually show that the brain doesn't really know where things are. Rather, it hypothesizes the probability of what the current event should look like and then, taking a leap of faith, approximates a viewable image. What you experience is not the image. What you experience is the leap of faith.

The brain does this because it needs to solve a problem: The world is three-dimensional, but light falls on the retina in a two-dimensional fashion. The brain must deal with this disparity if it is going to portray the world with any accuracy. To make sense of it all, the brain is forced to start guessing. Upon what does the brain base its guesses, at least in part? Experience with past events. After inserting numerous assumptions about the visual information (some of these assumptions may be inborn), the brain then offers up its findings for your perusal. Now you see one camel when there really is only one camel—and you see its proper depth and shape and size and even hints about whether it will bite you.

Far from being a camera, the brain is actively deconstructing the information given to it by the eyes, pushing it through a series of filters, and then reconstructing what it thinks it sees. Or what it thinks you should see. All of this happens in about the time it takes to blink your eyes. Indeed, it is happening right now. If you think the brain has to devote to vision a lot of its precious thinking resources, you are right on the money. Visual processing takes up about half of everything your brain does, in fact. This helps explain why professional wine tasters toss aside their taste buds so quickly in the thrall of visual stimuli. And why vision affects other senses, too.

Vision trumps touch, not just smell and taste

Amputees sometimes continue to experience the presence of their limb, even though the limb no longer exists. In some cases, the limb is perceived as frozen into a fixed position. Sometimes the person feels pain in the limb. Studies of people with phantom limbs demonstrate the powerful influence vision has on our other senses.

In one experiment, an amputee with a "frozen" phantom arm was seated at a table upon which had been placed a lidless box divided in half. The box had two portals in the front, one for the arm and one for the stump. The divider was a mirror on both sides. So the amputee could view a reflection of either his functioning hand or his stump. When the man looked down into the box, he could see his right arm present and his left arm missing. But when he looked at the reflection of his right arm in the mirror, he saw what looked like another arm. Suddenly, the phantom limb on the other side of the box "woke up." If he moved his normal hand while gazing at its reflection, he could feel his phantom move, too. And when he stopped moving his right arm, he felt his missing left arm stop also. The addition of visual information began convincing his brain of a miraculous rebirth of the absent limb.

A picture really is worth a thousand words

One way we can measure the dominance of vision is to look at its effect on learning. Researchers study this using two types of memory.

The first is called recognition memory, which underlies the concept of familiarity. We often deploy recognition memory when looking at old family photographs. Maybe you see a photo of an aunt not remembered for years. You don't necessarily recall her name, or the photo, but you still recognize her as your aunt. With recognition memory, you may not recall certain details surrounding whatever you see, but as soon as you see it, you know that you have seen it before.

The second involves working memory. Explained in greater detail in the Memory chapter, working memory is that collection of temporary storage buffers with fixed capacities and frustratingly short life spans. Visual short-term memory is the slice of that buffer dedicated to storing visual information. Most of us can hold the memory of about four objects at a time in that buffer, so it's a pretty small space. And it appears to be getting smaller. As the complexity of objects in our world increases, we are capable of remembering fewer objects over our lifetimes. Evidence also suggests that the number of objects and complexity of objects are engaged by different systems in the brain—turning the whole notion of short-term capacity, if you will forgive me, on its head. These limitations make it all the more remarkable that vision is probably the best single tool we have for learning anything.

When it comes to both recognition memory and working memory, pictures and text follow very different rules. Put simply, the more visual the input becomes, the more likely it is to be recognized—and recalled. It's called the pictorial superiority effect. Researchers have known about it for more than 100 years. (This is why we created a series of videos and animations of the Brain Rules at *www.brainrules.net*, making this book just one part of a multimedia project.)

The pictorial superiority effect is truly Olympian. Tests performed years ago showed that people could remember more than 2,500 pictures with at least 90 percent accuracy several days later, even though subjects saw each picture for about 10 seconds. (This is recognition memory, not working memory, at work.) Accuracy rates a year later still hovered around 63 percent. In one paper, picture recognition information was reliably retrieved several decades later. Sprinkled throughout these experiments were comparisons with text or oral presentations. The usual result was "picture demolishes them both." It still does. Text and oral presentations are not just less efficient than pictures for retaining certain types of information; they are *far* less efficient. If information is presented orally, people remember about 10 percent, tested 72 hours after exposure. That figure goes up to 65 percent if you add a picture.

Why is text less efficient than pictures? Because, it turns out, the brain sees words as lots of tiny pictures. A word is unreadable unless the brain can separately identify simple features in the letters. Instead of words, we see complex little art-museum masterpieces, with hundreds of features embedded in hundreds of letters. Like an art junkie, our brains linger at each feature, rigorously and independently verifying it before moving to the next. So reading creates a bottleneck in comprehension. To our cortex, surprisingly, there is no such thing as words.

That's not necessarily obvious. After all, the brain is as adaptive as Silly Putty. Given your years of reading books, writing email, and sending text messages, you might think your visual system could be trained to recognize common words without slogging through tedious additional steps of letter-feature recognition. But that is not what happens. No matter how experienced a reader you become, your brain will still stop and ponder the individual features of each letter you read—and do so until you can't read anymore.

By now, you can probably guess why this might be. Our evolutionary history was never dominated by books or email or text messages. It was dominated by trees and saber-toothed tigers. Vision means so much to us because most of the major threats to our lives in the savannah were apprehended visually. Ditto with most of our food supplies. Ditto with our perceptions of reproductive opportunity.

The tendency is so pervasive that, even when we read, most of us try to visualize what the text is telling us. "Words are only postage stamps delivering the object for you to unwrap," George Bernard Shaw was fond of saying. A lot of brain science now backs him up.

Vision is king from Day One

Babies come with a variety of preloaded software devoted to visual processing. We can determine what babies are paying attention to simply by watching them stare at their world. The importance of a baby's gazing behavior cannot be underestimated.

You can see this for yourself (if you have a baby nearby). Tie a ribbon around the baby's leg. Tie the other end to a mobile. At first she seems to be randomly moving her limbs. Soon, however, the infant learns that if she moves one leg, the mobile turns. She begins happily—and preferentially—moving that leg. Bring back the same mobile the next week, and the baby will move the same leg. Show the baby a different mobile, and she won't move the leg. That's what scientists found when they did this experiment. The baby is paying the most attention to the visual aspects of the mobiles. Since the mobiles don't look the same, there's not much reason to assume they would act the same. Babies use these visual cues even though nobody taught them to do so. This illustrates the importance of visual processing to our species.

Other evidence points to the same fact. Babies display a preference for patterns with high contrast. They seem to understand the principle of common fate: Objects that move together are perceived as part of the same object, such as stripes on a zebra. They can discriminate human faces from nonhuman equivalents and seem to prefer the human faces. They possess an understanding of size related to distance—that if an object is getting closer (and therefore getting bigger), it is still the same object. Babies can even categorize visual objects by common physical characteristics. The dominance of vision begins in the tiny world of infants.

It also shows up in the even tinier world of DNA. Our sense of smell and color vision are fighting each other for evolutionary control, for the right to be consulted first whenever something on the outside happens. And vision is winning. In fact, about 60 percent of our smell-related genes have been permanently damaged in this neural arbitrage, and they are marching toward obsolescence at a rate fourfold faster than any other species sampled. The reason for this decommissioning is simple: The visual cortex and the olfactory cortex take up a lot of neural real estate. In the crowded zero-sum world of the sub-scalp, something has to give. Does this mean that we'll permanently lose our sense of smell or that our heads are no longer getting bigger? Check back in several hundred thousand years. The evolutionary forces that actively selected against smell are not still in full force today. But what forces are replacing them is an active area of debate.

Whether looking at behavior, cells, or genes, we can observe how important the visual sense is to the human experience. Striding across our brain like an out-of-control superpower, giant swaths of biological resources are consumed by it. In return, our visual system creates movies, generates hallucinations, and consults with previous information before allowing us to see the outside. It happily bends the information from other senses to do its bidding and, at least in the case of smell, seems to be caught in the act of taking over.

When it comes to applying this knowledge in your own daily life, is there any point in trying to ignore the vision juggernaut? You don't have to look any further than the wine experts of Bordeaux for the answer.

More ideas

The best visuals for learning

What kind of pictures best grab attention and thus transfer information? We pay lots of attention to color. We pay lots of attention to orientation. We pay lots of attention to size. And we pay special attention if the object is in motion. Indeed, most of the things that threatened us in the Serengeti *moved*, and the brain has evolved unbelievably sophisticated trip wires to detect motion. We even have specialized regions to distinguish when our eyes are moving versus when our world is moving. These regions routinely shut down perceptions of eye movement in favor of the environmental movement.

That said, we need more research into practical applications. The pictorial superiority effect is a well-established fact for certain types of classroom material, but not for all material. Data are sparse. Do pictures communicate conceptual ideas such as "freedom" and "amount" better than, say, a narrative? Are language arts better represented in picture form or using other media? It's unclear.

Include video or animation

I owe my career choice to Donald Duck. I am not joking. I even remember the moment he convinced me. I was 8 years old at the time, and my mother trundled the family off to a showing of an amazing 27-minute animated short called *Donald in Mathmagic Land*. Using visual imagery, a wicked sense of humor, and the wide-eyed wonder of an infant, Donald Duck introduced me to math. Got me excited about it. From geometry to football to playing billiards, the power and beauty of mathematics were made so real for this nerd-in-training, I asked if I could see it a second time. My mother obliged, and the effect was so memorable, it eventually influenced my career choice. I now have a copy of those valuable 27 minutes in my own home and regularly inflict it upon my poor children. *Donald in Mathmagic Land* won an Academy Award for best animated short of 1959. It also should have gotten a Teacher of the Year award. The film illustrates—literally—the power of the moving image in communicating complex information to students.

Animating presentations is another way to capture the importance not only of color and placement but also of motion. The basics are not hard to learn. With today's software, anybody who knows how to draw a square and a circle can create simple animations. Simple two-dimensional pictures are quite adequate; studies show that if the drawings are too complex or lifelike, they can distract from the transfer of information.

Communicate with pictures more than words

"Less text, more pictures" were almost fighting words in 1982. They were used derisively to greet the arrival of *USA Today*, a brand-new type of newspaper with, as you know, less text, more pictures. Some predicted the style would never work. Others predicted that if it did, the style would spell the end of Western civilization as the newspaper-reading public knows it. The jury may be out on the latter prediction, but the former has a powerful and embarrassing verdict. Within four years, *USA Today* had the second-highest readership of any newspaper in the country, and within 10 years, it was number one. It still is.

What happened? Pictorial information may be initially more attractive to consumers, in part because it takes less effort to comprehend. Because it is also a more efficient way to glue information to a neuron, there may be strong reasons for entire marketing departments to think seriously about making pictorial presentations their primary way of transferring information.

The initial effect of pictures on attention has been tested. Using infrared eye-tracking technology, 3,600 consumers were tested on 1,363 print advertisements. The conclusion? Pictorial information was superior in capturing attention—independent of its size. Even if the picture was small and crowded with lots of other nonpictorial elements close to it, the eye went to the visual.

Toss your PowerPoint presentations

The presentation software called PowerPoint has become ubiquitous, from work meetings to college classrooms to conferences. What's wrong with that? They're mostly text, even though they don't have to be. A typical PowerPoint business presentation has nearly 40 words *per slide*. Please, do two things: (1) Burn your current PowerPoint presentations. (2) Make new ones. Then see which one works better.

Brain Rule #9
Vision trumps all other senses.

- Vision is by far our most dominant sense, taking up half of our brain's resources.

- What we see is only what our brain tells us we see, and it's not 100 percent accurate.

- The visual analysis we do has many steps. The retina assembles photons into movie-like streams of information. The visual cortex processes these streams: some areas registering motion, others registering color, etc. Finally, we recombine that information so that we can see.

- We learn and remember best through pictures, not through written or spoken words.

music

Study or listen
to boost cognition.

HENRY DRYER IS A 92-year-old dementia patient living in an assisted living center. Henry sits alone in a wheelchair in the middle of a room, eyes downcast, face empty. His body seems vacant too. In the documentary film featuring him, Henry is described by famed neurologist Oliver Sacks as "inert, maybe depressed, unresponsive, and almost unalive." Henry has barely spoken to anyone in the decade he's lived at the center. This is not how he used to be, his daughter relates. Henry was outgoing for most of his life, blessed with a passionate love affair for the Bible and for dancing and singing. It was not unusual for him to spontaneously burst out into song in public.

On this day, Henry is part of a project helping elderly people reconnect by listening to music they love. Henry is given an iPod loaded with music. As soon as Henry hears the music, Henry starts making a noise like a horn. Suddenly, Henry's eyes grow wide. His face instantly lights up, a bit contorted. Henry grabs his wrists and starts swaying, smiling, and singing. Henry becomes *alive*.

When the iPod is turned off, Henry doesn't slink back into silence. He becomes articulate, funny and *very* enthusiastic. "Do you like music?" someone asks off-camera. Henry answers, "I'm CRAZY about music. You play beautiful music. Beautiful sounds!" "What was your favorite music when you were young?" Cab Calloway, Henry responds, then starts scatting. He sings, "I'll Be Home for Christmas" with accurate pitch, wonderful emotion, and occasionally correct lyrics.

He is asked "What does music do to you?" Face still animated, arms now gesticulating with purpose, Henry responds: "It gives me the feeling of love. Romance! I figure right now the world needs to come into music, singing, you've got beautiful music here. Beautiful. Lovely. I feel a band of love!"

Dr. Sacks is delighted. "In some sense Henry is restored to himself," he enthuses. "He has remembered who he is, and he's reacquired his identity for a while through the power of music." I barely heard Dr. Sacks, because I started tearing up. It's one of the most moving videos I've ever seen.

How does music light up the brain, as it clearly did for Henry? What effects does it have on young and old? What does listening to music do to the brain, compared with being trained in music? Scientists have intensively investigated these questions. In asking whether exposure to music produces benefits in nonmusical cognitive domains, scientists have looked at academic areas, like reading and math. They've looked at general intelligence. They've studied the effects of music on speech, physical development, and mood. And now we think we have an understanding of at least some of the effects of music on cognition.

Why "think" instead of "know"? Music research is complicated—starting with the fact that not everyone agrees what music is, or why it exists.

How would you define music?

Scientists aren't sure how the brain defines music, in part because there is no universal agreement about exactly what music is. What may be annoying, unorganized, environmental noise to a person raised in culture A at time point A might be rapturous, organized, beautiful *music* to a person raised in culture B at time point B. For example, in 1971, George Harrison of The Beatles organized a benefit concert, called The Concert for Bangladesh, with sitar master Ravi Shankar. Shankar tuned his instrument before performing, an event heard over the loudspeakers by the mostly Western audience. The crowd clapped and cheered with wild enthusiasm. As they began to settle, Ravi addressed them: "Thank you. If you like our tuning so much, I hope you will enjoy the playing more." Rap is another example. It is clearly speech and also clearly—what? Music? Generations don't agree. Neither do composers. Neither do sociologists. One professor of music and science at Cambridge defines music this way: "Musics (yes, the author said musics) can be defined as those temporally patterned activities, individual and social, that involve the production and perception of sound and have no evident and immediate efficacy or fixed consensual reference." That's not exactly the way everyone would describe music. The definition of music has been so tough to determine that neuroscientist Seth Horowitz, in his book *The Universal Sense*, titled a chapter "Ten Dollars to the First Person who Can Define 'Music' (and Get a Musician, a Psychologist, a Composer, a Neuroscientist and Someone Listening to an iPod to Agree . . .)."

And yet, at some level, we all know what music is, as did our ancestors. Music has tempo, changes in frequency, and something we call timbre (the quality that separates the "sound" of a sitar from the "sound" of a violin, for example). It is often associated with movement, such as dancing. It is a real phenomenon, even if it is elusive to define.

Some scientists think we are born musical. You can certainly watch babies respond to music, swaying and responding with glee to specific intervals. They even love it when parents talk to them in musical speech called "parentese," which is rhythmic and high-pitched, with long, drawn-out vowels. Music has been a part of the cultural expression of virtually every culture ever studied. It may even extend into prehistoric times. A 35,000-year-old flute made from bird bone has been discovered, to cite just one example. If every culture has some form of musical expression, and if babies so readily respond to it, some scientists say, music must serve some evolutionary function. We must be hardwired for music, with regions in the brain specifically devoted to music.

Harvard professor Steven Pinker begs to differ. "I suspect that music is auditory cheesecake, an exquisite confection crafted to tickle the sensitive spots of at least six of our mental faculties," he writes in *How the Mind Works*. Like music, people love cheesecake, and they have for a very long time (a recipe for cheesecake was found around 5th century BCE). But that doesn't mean the brain has a region specifically dedicated to cheesecake. We are hardwired to respond not to cheesecake specifically, Pinker says, but to fats and sugars. These major energy boosters were somewhat rare in the lean world of the Serengeti. Because of their scarcity, our brains became sensitized—dedicated, you might say—to detecting the presence of fats and sugars. Because of their value, our brains rewarded their consumption with a powerful jolt of pleasure. Pinker makes a similar argument for music. He thinks music stimulates specific regions in the brain that are actually hardwired to process nonmusical inputs. There is no reason to go after evolutionary arguments that explain dedicated musical modules in the brain, Pinker posits, for a very practical reason: there are none.

So the matter is unsettled on why music exists, and scientists don't agree on how to even define music. Still, researchers forge ahead with studies on cognition and social skills. They've discovered fascinating ways that music may benefit the brain. The benefits just aren't the ones that the average person thinks they are.

What music training does for the brain

Ray Vizcarra was an award-winning music and band teacher in a Los Angeles high school. He took kids who had no musical training, and he whipped them into shape with such skill, and such speed, that the kids were soon winning all-city contests. That's saying something, given that Los Angeles is ground zero for musical contests. The LA City Council singled him out for a special honor in 2011. And then he lost his job. In a round of budget cuts and layoffs, he didn't have enough seniority to stay. The story was written up in the *Los Angeles Times*.

Most of my wife's friends are professional musicians, and they were outraged. They saw his layoff as one more sad example of music falling to the wayside now that schools emphasize standardized tests, which favor reading and math. Invariably, the conversation turned to questions about the value of keeping music in schools. Doesn't music help improve test scores in reading and math? they ask me.

My response is not what they expect.

"It's not a simple story," I usually respond. Then I start listing the variables. When they say "music," do they mean listening to music all the time? Or do they mean music training, like what the band teacher did with his students? Both involve exposure to music but are hardly the same thing. Does "help" mean changing an SAT score? How about cognitive processes not generally covered by standardized tests; do those count?

Usually they're talking about the effect of music lessons on reading ability, math scores, or intelligence in general. And in that case, I have bad news—made worse because I first need to spend a few minutes giving a statistics lesson. The lesson centers around something called an *r* value.

An *r* value is a quantifiable linear association between two variables. It measures the tightness of their relationship. *R* values are assigned a number between -1 and 1. As an *r* value gets closer to 1, there is an increasingly positive relationship between the two variables. My wife, to give one example, loves chocolate. Every time she eats it, she breaks out into a big smile. The relationship between chocolate and smile is tight. We could easily assign it an *r* value of 1.

In science, we use *r* values when reviewing multiple investigations done over a period of years to look for patterns—called a meta-analysis. That's usually the kind of study done to analyze whether music is associated with a boost in academic or cognitive performance. Let's look at the actual results in a few areas rumored to be true.

Music training improves math scores. The best score in the literature gives the association an *r* value of 0.16. That's not much.

Music training improves reading ability. This sports an *r* value of about 0.11. In more recent studies, researchers are beginning to detect improvement in reading skills of musicians compared to non-musicians, but more research is needed.

Music training improves IQ. The answer again is no. Musicians *are* smarter, but the reason may be that smarter people take music lessons.

Music training improves something useful for academics, right? Yes: spatiotemporal reasoning. That's the kind of reasoning that allows you to, among other things, rotate three-dimensional images in your head. This is the kind of skill used by an architect or engineer. There's an *r* value of 0.32 between the two if you take group instruction in piano, 0.48 if you take individual lessons.

This is not an impressive track record, taken together.

Nonetheless, *r* values even lower than these can make headlines. One of my favorite examples is the so-called Mozart Effect. Listening to Mozart, the news stories claimed, will improve your ability to do math. An entire cottage industry grew up around this phenomenon, selling DVDs and CDs marinated in Mozart, then marketed to anxious parents worried about their child's cognitive development. At one point, the governor of Georgia issued classical music CDs to the parents of every newborn in the state. The basis of all this enthusiasm was a tiny little paper that got a giant dollop of publicity because it was published in the prestigious journal *Nature*. The paper showed that when undergraduate students listened to 10 minutes of Mozart just before taking spatial tests, their scores improved. The boost was not strong, and the statistical analysis was even less so. The r value was a miserable 0.06. *Nature* issued a critique of the paper a month later, questioning the finding. Scientists who tried to replicate the results found that *any* pleasurable listening (or reading) experience had the same effect—one lasting about 15 minutes. But that not-so-shiny fact generated almost no publicity. The lead author of the original study has denounced the cottage industry, and years later reflected that the money Georgia's governor appropriated for the music CDs might have been better spent on music education in the public schools.

That study was published more than 20 years ago. But even when I lecture on brain science today, I encounter people who think classical music is good for your brain. Happily, music does do the brain some good. First we'll look at the effects of taking music lessons, and then the effects of listening to music.

Musicians are better listeners

Let's say you are in a lab listening to some audio that is familiar and predictable. All of a sudden the scientist inserts some change into the sound you are hearing (a rhythmic pattern change, or a pitch change, for example). This alteration could be dramatic or subtle, but the scientist is interested in one question: Did you detect it? The more subtle the change you can detect, the higher your score is.

Musicians score better than nonmusicians on such tests. But here's the interesting thing. They also score better when the audio being played is speech, not music. For example, musicians show more robust neurological stimulation than nonmusicians to the frequency changes of their native tongue. Musicians also are better able to pick out and pay attention to a specific sound in a roomful of distracting noises. (The fancy name for this is auditory stream segregation.)

Music training boosts language skills

In one study, researchers gave children twice-weekly music lessons for a school year, using "a music curriculum designed to teach prereading and writing skills." The children's neuroarchitecture changed in a way that boosts both motor skills (writing) and auditory skills (word recognition)—direct improvements in language processing. Ten-year-olds who have been practicing a musical instrument for at least three years see a boost in both their vocabulary and nonverbal reasoning skills over children who don't. Kids who start music lessons prior to first grade show superior sensory-motor integration when they are adults. These findings alone make a strong case for parents starting music lessons before age 7.

Musical training provides direct improvements in working memory, not only in the phonological loop but also in the visuospatial sketch pad (see the Miguel Najdorf story in the Memory chapter for more on that). Working memory is a key constituent of executive function. Executive function predicts students' future undergraduate performances better than their SAT scores, or even their IQs. Selecting and focusing on relevant stimuli from a host of choices is also a component of executive function. Any assistance music provides in this domain (and helping students pick out specific auditory streams in a room filled with irrelevant noise is one big example) is probably a good thing for kids.

Taken together, these studies make a case for supporting music education. In the journal *Nature Reviews Neuroscience*, researchers Nina Kraus and Bharath Chandrasekaran write of the studies on listening: "The beneficial effects of music training on sensory processing confer advantages beyond music processing itself. This argues for an improvement in the quality and quantity of music training in schools."

Music to Ray Vizcarra's ears, no doubt.

The link between speech and music

Why would music training benefit speech? We know that music and speech are not processed identically in the human brain. But we also know they share many common features.

Take rhythm, for one. People can speak in a pulsed pattern, as when reading a Shakespearean play, or a poem, or a rap. As any drummer will tell you, rhythm is very much a part of the musical experience, too.

Take pitch, for another. When people are finished speaking a sentence, the pitch of their voice invariably lowers. When people ask a question, their voice invariably rises. Pitch variation is a key part of speech. It is also one of the signature hallmarks of music.

Music processing in the brain may, I believe, be conceptually likened to a Venn diagram, where two circles partially overlap to create a shared region. The brain has regions that are speech-specific. Call it the red domain. And the brain has regions that are music-specific. Call it the blue domain. But speech and music also share some regions in common—psychologically and physiologically. With apologies to Alice Walker, color it purple.

The brain keeps its separate regions quite separate, as we know from cases like Monica, a Canadian nurse who suffers from a condition called congenital amusia. Monica can't carry a tune in a bucket. Neither can many members of her family. Her condition, however, is not just that she can't match the pitch she hears in a song. Studies show that Monica cannot discriminate between notes. She literally can't tell one note from the other, can't determine if one is "sour" compared to another, can't detect melodic patterns of any kind. With respect to music, she is completely tone-deaf. Monica does not enjoy listening to music. It appears to be a source of stress, as perhaps her schoolmates could attest: Monica was in her church choir and school band as a little girl.

You would never know that Monica has pitch discrimination issues if you struck up a conversation with her, however. She speaks just like the rest of us. Her voice goes down when she finishes a declarative sentence (she's no Valley Girl), and her voice goes up when she is finished with a question. Monica can detect these changes in pitch, in both her voice and the voice of anyone else.

In another case of amusia, a child attempted piano lessons. His instructor soon found he could not discriminate between two pitches (and also could not keep time). When it came to speech, though, it was a different story. He fluently spoke three languages besides his native tongue.

It seems odd that people can detect pitch changes when their brains decide they are listening to *speech*, but they become completely addled if their brains decide they are listening to *music*. When sound waves enter your ear, how does the brain determine whether you are listening to environmental noise, speech, or music? This question turns out to be important for a variety of reasons. As we shall see later, people who have lost speech abilities can often regain them through exposure to music. That doesn't happen if all they hear is the spoken word. How does that work? What is the brain's criteria for distinguishing music? Scientists don't know. We just know that the brain at some point seems to separate music from speech.

However, it's the purple section of our Venn diagram—the area where the neurological processing domains for speech and music overlap—that is most interesting to the question at hand. This overlap is the reason that music training affects aspects of speech: if you improve one, you can also improve the other.

Music lessons improve social skills

What else can music training do—besides, of course, make people better musicians? Watch the jazz band the Pat Metheny Group play "Have You Heard" live, and you may get an idea.

Pat Metheny is a bushy-haired American jazz guitarist and composer, winner of 19 Grammy Awards. He has been making records since the mid-1970s. I saw a video of him performing live in Japan in 1995, and the group's improvisatory prowess was on full display. Besides the joyous virtuoso performance, the impression that strikes me most is the almost ridiculous cooperation of the band. There are five saxes, five trumpets, two vocalists, a string bass, keyboards, several rhythm sections, and probably a bunch of people I can't see. There is plenty of room for error, yet that is exactly what you don't hear. The musicians switch off performing solos throughout the song, tossing around melodies like Frisbees, and yet they play as one person. They don't even have to look at each other—they can't, in fact; the stage is mostly dark. The musicians signal to each other using the subtle nonverbal cues so legendary in jazz performance, creating musical dialogues only seasoned musicians can make intelligible. It is exhilarating, magical stuff.

How do they achieve such coordination? Is there something about performing in a musical group that trains people to look for subtle cues in others, in the service of coordinating a goal-oriented activity? Behavior done for the good of a group, or for the good of another individual, is termed "prosocial." The action could be as exotic as allowing another musician solo space in a jazz concert so that he or she may shine, or it could be as mundane as making dinner when your spouse is sick. Prosocial skills, you can imagine, profoundly influence a person's social abilities in all aspects of life.

Does music training confer social, not just cognitive, benefits? You don't have to be good enough to play in Mr. Metheny's band to know that is exactly what one finds. The research we'll look at next spans the age spectrum from adults to infants.

Musicians are better at detecting emotion

If you've ever cried because you were yelled at, you know: words convey emotions. You can find out *what* somebody is feeling by detecting *how* they are saying something. We call such abilities "vocal affective discrimination skills." Researchers asked: How good are trained musicians at these skills compared to nonmusicians?

In one study, English-speaking musicians and nonmusicians heard various emotions expressed in Tagalog, a Philippine language that was foreign to them. They were asked to identify any emotion they heard. How good were they at detecting the emotional information in what was being said, even though they could not understand the words? The results were dramatic. Trained musicians were champs, while nonmusicians were surprisingly bad at it. Musicians were especially good at discerning sadness and fear. They actually scored higher when listening to Tagalog than when listening to their native English! Studies like these laid the groundwork for demonstrating that music might improve social skills.

Another research effort involved college-age students who had received musical training for 10-plus years. The researchers eavesdropped on the students' brain activity using noninvasive imaging technologies while playing various auditory cues. They were specifically interested in the students' brain stems—the primal, most evolutionarily ancient parts of our brains. What exactly were their brains doing as they listened to the audio cues, compared to the brains of nonmusicians?

Consistent with previous findings, the researchers found that the musicians outpaced the nonmusicians in discriminating emotional information. These undergraduates were especially good at detecting subtle changes in the sound, timing, and pitch of a *baby's* cry, for heaven's sake. (Getting this right can be enormously difficult to do.) We call such talents fine-grained discrimination. Extending the previous findings, the researchers showed that musicians' brain stems were more efficient at this neural-processing task. Specifically, their brains exhibited increased time-domain responses to complex emotional information. Their *brains*, not just their behaviors, were better.

Much research remains to be done, however. It's unclear whether music training directly improves this ability, or whether people who are naturally better at fine-grained discrimination have a tendency to like music and stick with music lessons.

Music lessons make kids more empathetic

Researchers wanted to know whether music training could directly *cause* changes in social ability.

Fifty kids, ages 8 to 11, were randomly assigned to one of three groups. The first group took group music classes for an entire academic year. The delightful curriculum consisted of rhythmic improvisation, musical games, melodic repetition, and shared musical experiences. The second group played games that also involved imitating and interactive experiences—but verbal mostly, no music. The third group simply attended the regular school year. The question was: How good were the children's social abilities at the end of the school year? Before the experiments commenced, researchers established baseline measures by testing the children's social skills, such as empathy, including Theory of Mind abilities.

The children in the music group had the most improved empathy scores. Like the adults, these kids had a stronger ability to decode the emotional information in their social surroundings, both verbally and nonverbally. They also were better at imitating facial expressions. The children who took the music class also had more empathetic responses to artificially posed situations, as measured by the Bryant's Index of Empathy (an instrument used to measure pediatric empathy). The other two groups showed no such improvement.

Said lead researcher Tal-Chen Rabinowitch, "Overall, the capacity for empathy in children that participated in our musical group interaction program significantly increased."

The experiment has since been replicated with 6-year-olds, by researchers in Canada.

Infants are more social, too

So far, we can detect the social benefits of music lessons in older adults, undergraduates, and elementary-school children. How far back can you push this? Can you detect social benefits if you give music lessons to *infants*? You can't go much earlier than that. Amazingly, the researchers found similar findings.

Six-month-old babies took a parent-and-child music class for six months. The instruction was based roughly on Suzuki methodology, one that requires active group participation. Activities involved lots of singing, lots of banging on instruments, and learning songs in class, which parents were asked to repeat at home. Not surprisingly, this group was called the Active Group. A second group served as the control. These parents and tots instead listened to *Baby Einstein* music CDs while playing with toys together. Predictably, they were called the Passive Group.

You can actually measure social competence in babies using a complex instrument called the Infant Behavior Questionnaire (IBQ), which assesses infants on 14 aspects of temperament. Researchers measured both groups to get a baseline. Then the experiment commenced. How did the babies do? If you are a music advocate, get ready for some spine-tingling data.

The Active Group outpaced the Passive Group socially in virtually every way you can measure it. They smiled more. They laughed more. They were much easier to calm down when they were stressed. In limitation assessments (a measure of how well you react to unexpected stimuli), they exhibited much less stress than their Passive counterparts. The infants' gestures—such as waving good-bye and pointing—were improved, a companion paper showed. That may be important. Such prelinguistic communication leads to more positive social interactions between parent and child. And *that* improves infant cognition in virtually every way you can measure it.

What's going on here? We don't know for sure. The Passive Group was exposed to the same amount of music as the Active Group, as well as the same amount of social interaction. Making music may simply provide an environment where one gets to exercise greater social cooperation and generally prosocial behaviors than when playing with toys. In this view, the secret sauce lies not with the music, but with the interaction. Or it could be the music itself, for both groups of children experienced sustained interaction with their parents. Either way, a method involving music has been found to make kids more empathetic, more relational.

Which is the point.

Though these and several other experiments are interventions, showing whether music training directly caused the effects, the vast majority of studies are associative in nature. Still, taken together, these studies suggest—sometimes strongly—that music training boosts foundational speech-processing tasks, spatial skills, the detection of emotional cues, empathy, and baby-size social skills. Next, let's look at the effects of simply listening to music.

Music changes your mood

"The word is *breast!*" my mother yelled from the kitchen. This brought my 13-year-old mind very quickly to attention. She clarified: "Music soothes the savage *breast!* I believe it was from some old play …" her voice trailed off.

I was in the TV room, watching a Bugs Bunny cartoon called *Hurdy-Gurdy Hare*, and my mother had overheard a line. The plot was standard Looney Tunes fare, with dollops of humor for both adults and children, involving an escaped gorilla now after Bugs. After a lot of antics, the gorilla traps Bugs Bunny in the back room of an apartment. Conveniently, and in the nick of time, Bugs finds a violin and begins playing. Immediately the gorilla calms down, then begins moving to the music. Bugs says snarkily says to the camera, "They say music calms the savage beast." I did not see what happened next because of my mom's comment. She was right, of course. According to scholars, the line is from the pen of 17th-century playwright William Congreve, and properly reads "Music hath charms to soothe the savage breast."

Either way, music's ability to affect one's mood and subsequent behavior is a common theme in literature. Researchers will tell you the reason is biochemical. It is a surprisingly well-established fact that music can induce hormonal changes. These changes result in alterations of mood. Well duh, say music fans around the world. Anybody who has ever listened to their favorite song could testify to *that*. It is not earth-shattering to find that music can induce pleasure. "Enjoyment arousal," as it's called, is sometimes accompanied by a temporary boost in certain skills. For that, we can thank three hormones: dopamine, cortisol, and oxytocin.

Dopamine

Noted Canadian researcher Robert Zatorre has studied people's emotional reactions to music for a long time. He and his colleagues have found that when people hear their very favorite music (I mean spine-tingling, awe-inspiring, fly-me-to-the-moon music), their bodies dump dopamine into a specific part of their brain.

Dopamine is a neurotransmitter, involved in mediating processes from feeling pleasure to memory formation. It floods the striatal system, a curved structure in the middle of the brain that's involved in many functions, including evaluating the significance you assign to a given stimulus. Zatorre found that when you hear music that gives you goose bumps (called "musical frisson"), the striatal system is activated via dopamine release. Music may soothe the savage human by exploiting this mechansim.

Cortisol

Surgery is not a pleasurable experience for most people. Some patients are genuinely freaked out, however, to the point of requiring medical intervention. Researchers asked, "Could music reduce the stress of people about to undergo surgery?" To answer the question, they divided 372 patients into two groups. The first group would listen to music before going under the knife. The second group would take an antistress pill (midazolam) prior to surgery.

Who experienced the least amount of stress, as measured by respiration and heart rate, among other assays? The music group. They felt 13 percent less anxious than the stress-pill group before their surgeries. Listening to classical or meditation music had the greatest effect.

Oxytocin

Oxytocin plays a huge role in social bonding. This talented molecule stimulates temporary feelings of trust, orgasms, lactation, and even birth (pitocin, a drug that induces contractions, is a synthetic form of oxytocin). It even gets some mammals, like the prairie vole, to mate for life. Given this social track record, it is a big deal when the brain when increases its production of oxytocin as a response to some external cue.

Researchers have discovered that when people sing as a group, as they would in a choir, oxytocin courses through their brains. An uptick in the hormone is a fairly reliable indicator of feelings of trust, love, and acceptance. This may explain why people in a choir often report feeling so close to each other.

University of Montreal researcher Dan Levitin, in an interview with NPR, said the same of playing music together: "We now know that when people play music together, oxytocin is released.... This is the bonding hormone that's released when people have an orgasm together. And so you have to ask yourself, that can't be a coincidence;

there had to be some evolutionary pressure there. Language doesn't produce it, music does...." This flies in the face of Pinker's auditory cheesecake, as you may have noted.

These data suggest a mechanism whereby music makes people happy, calms them down, maybe even makes them feel close to each other. I can personally attest to these feelings.

My wife is a classically trained pianist and a composer (she scores documentaries). In the past few years, she has really gotten into Irish, Scottish, and Celtic music. One gorgeous Gaelic song she regularly listens to speaks to me also. I'm hydrated with this glorious cocktail of haunting, calming, restful feelings, right from its opening bars. That turned out to be important on a day we had driven from Seattle to Vancouver, British Columbia. We were on vacation, and I was not having a restful time at all. It was downtown at rush hour—Vancouver at its worst—and I was in a slow burn trying to find our hotel, my tension increasing with every missed intersection. Stress hormones were boiling my blood, something my wife is good at detecting. She found the CD with that Gaelic song, slipped it into the car stereo, and played it full volume. From a distance I detected the calming feelings. I attempted to give in to them and immediately felt peace wash over me. We quickly found our lodgings. As I can attest, the calming ability of music can be very pleasurable ... especially for the other people in the car.

But more importantly, these hormones represent a powerful effort from researchers to transform anecdotal, ephemeral impressions about the power of music into the exacting physical world of cells and molecules. The findings may have medical implications.

The promise of music therapy

Using music as medicine for sick patients has a long history. The Greek physician Hippocrates prescribed it for mentally ill patients. During World War I, hospitals in the UK employed musicians to play for wounded soldiers in convalescence. It seemed not only to calm them down but also to reduce their pain. None of this was measured in any formal way at the time, but the observation was so persistent that the practice continued into World War II. Observations like these eventually led to the establishment of formal music-therapy associations.

Slowly but surely, these anecdotal observations attracted the notice of the research community, and clear findings have emerged. Music has been shown to aid speech recovery in head-trauma patients, for example. Gabrielle Giffords (the US representative who survived a gunshot to the head) regained regular speech in part by singing. Researchers think it works by forcing the brain to sign up unused regions of the brain for speech duty. Nobody knows why music does this. Dr. Oliver Sacks, interviewed about Giffords's recovery in a documentary, said: "Nothing activates the brain so extensively as music. It has been possible to create a new language area in the right hemisphere. And that blew my mind."

Music improves the recovery rates of specific cognitive abilities in stroke patients. In one study, patients who underwent six months of music therapy were compared to patients who got "talk therapy." The results were extraordinary. In measurements of verbal memory, the talk therapy patients achieved a score of 7 (that's not so good). The music group achieved a score of 23 (that's really good). Measurements of focused attention showed a similar disparity: the talk-therapy group scored a 1, while the music-therapy group scored an 11. In overall language skills at the end of six months, the talk-therapy group scored a 5. The music-therapy group scored a 21.

Among stroke patients with motor difficulties, including those with Parkinson's and cerebral palsy, researchers find similar positive results. Music-therapy patients routinely outscore patients exposed to more traditional therapies in measurements of arm movements and of gait as they walk. Music seems to serve as a predictable metronome that helps people coordinate their movements.

Most of these studies have been done on adults, often our oldest citizens. What about some of our youngest?

Prematurely born infants, living in a hospital's Neonatal Intensive Care Unit (NICU), gained weight more rapidly when music was played. Music helped them learn how to suck at their mothers' breasts more readily. It also reduced their overall stress levels, which may explain the other findings. One study found that female (though not male) infants' stay in the unit would be decreased by 11 days if music were played, compared to no music. It is now standard for hospitals across the country to pipe calm, peaceful music into their NICUs.

Why does music have these effects? Again, we don't know for sure. One idea, the "arousal and mood hypothesis," was published in 2001. It proposed that the three hormones explain why music speeds recovery. It's still just a hypothesis, but it's paving the way for some serious neuroscience. Stay tuned.

More ideas

Too many of these intriguing studies don't *prove* cause, and they're all done in a lab setting. I'd like to see a school district take up research on music programs and help determine the effects of music training in a real-world setting. As soon as kids enter first grade, schools would randomly assign a large number of them to one of two groups. The first group would take lessons on a musical instrument, with formal instruction and ensemble training. Lessons would be daily, consistent, and as mandatory as math class. The program would last at least 10 years, ending when the students are juniors in high school. The second group would receive no music training.

With this kind of large-scale, long-term research program, we could see whether students who get music training perform better on tests involving speech proficiency at the end of the 10-year period than those without the training. And language arts. And second languages. Since emotional regulation has such a powerful effect on academic performance (see the Stress chapter), additional questions are relevant as well. We could see if the kids with music training have better emotional regulation. If they get better grades. If they're more cooperative in group settings not related to music. If music training reduces antisocial behavior, such as bullying, at school. Music training almost certainly teaches discipline, a form of impulse control (you continue practicing for 10 years, even if you'd rather not).

If the answer was affirmative to even one of these questions, we would end up with a truly interesting principle: One way to create a higher-functioning student is to hire back band teacher Ray Vizcarra. And if it comes time to cut the school budget, the last activity to go would be formal musical training.

Brain Rule #10
Study or listen to boost cognition.

* Formal musical training improves intellectual skills in several cognitive domains. Music boosts spatiotemporal skills, vocabulary, picking out sounds in a noisy environment, working memory, and sensory-motor skills.

* Formal music training also aids social cognition. People with music training are better able to detect the emotional information in speech. Empathy skills and other prosocial behaviors improve.

* Variations on these effects have been shown in adults, college students, schoolchildren, even infants.

gender

Male and female brains are different.

THE MAN WAS A hot dog. The woman was a bitch.

The results of the experiment could be summarized in those two sentences. Researchers had asked people to rate a fictional person's job performance—an assistant vice president of an aircraft company. People were divided into four groups, each group with an equal number of men and women, for the experiment. The groups were given the vice president's brief job description. But the first group also was told that the vice president was a man. Asked to rate both the competence and the likability of the candidate, this group gave a very flattering review, rating the man "very competent" and "likable." The second group was told that the vice president was a woman. She was rated "likable" but "not very competent." All other factors were equal. Only the perceived gender had changed.

The third group was told that the vice president was a male superstar, a stellar performer on the fast track at the company. The fourth group was told that the vice president was a female superstar, also on the express lane to the executive washroom. As before, the third group rated the man "very competent" and "likable." The woman superstar also was rated "very competent." But she was not rated "likable." In fact, the group's descriptions included words such as "hostile." As I said, the man was a hot dog. The woman was a bitch.

The point is, gender biases hurt real people in real-world situations. As we hurtle headlong into the controversial world of gender differences, keeping these social effects in mind is excruciatingly important. There is a great deal of confusion regarding the way men and women relate to each other, and even more confusion about why they relate to each other differently. Terms are often confused as well, blurring the line between the concepts of "sex" and "gender." In this chapter, sex will generally refer to biology and anatomy. Gender will refer mostly to social expectations. Sex is set into the concrete of DNA. Gender is not.

Differences between men's and women's brains can be viewed from several lenses: genetic, neuroanatomical, and behavioral. Scientists usually spend their whole careers exploring only one. So our tour of all three will be necessarily brief.

How we become male or female

The differences between men's and women's brains start with genes, which determine whether we become male or female in the first place. The road to sex assignment starts out with all the enthusiasm sex usually stimulates. Four hundred million sperm fall all over themselves attempting to find one egg during intercourse. The task is not all that difficult. In the microscopic world of human fertilization, the egg is the size of the Death Star, and the sperm are the size of X-wing starfighters.

X is the name of that very important chromosome that half of all sperm and all eggs carry. You recall chromosomes from biology class. They're those writhing strings of DNA packed into the cell nucleus that contain the information necessary to make you. You can think of chromosomes as volumes in an encyclopedia. Creating you takes 46 of them. Twenty-three come from Mom, and 23 come from Dad. Two are sex chromosomes, either X or Y. At least one of your sex chromosomes has to be an X chromosome, or you will die. If you get two X chromosomes, you go into the ladies locker room all your life; an X and Y puts you forever in the men's. The Y can be donated only by sperm—the egg never carries one—so sex assignment is controlled by the man. (Henry VIII's wives wish that he had known that. He executed one of them, Anne Boleyn, for being unable to produce a son as heir to the throne, but it would have made more sense to execute himself.)

What the X and Y chromosomes do

One of the most interesting facts about the Y chromosome is that you don't need most of it to make a male. All it takes to kick-start the male developmental program is a small snippet near the middle, carrying a gene called SRY.

David C. Page is the researcher who isolated SRY. Though in his 50s, Page looks to be about 28 years old. As director of the Whitehead Institute and a professor at MIT, he is a man of considerable intellect. He also is charming, with a refreshingly wicked sense of humor. Page is the world's first molecular sex therapist. Or, better, sex broker. He discovered that you can destroy the SRY gene in a male embryo and get a female, or add SRY to a female embryo and turn her into a male (SR stands for "sex reversal").

Why can you do this? In a fact troubling to anybody who believes males are biologically hardwired to dominate the planet, researchers discovered that the basic default setting of the mammalian embryo is to become female. Yet the male program is enthusiastic. The CIA estimates (though not everyone agrees) that 107 male babies are born for every 100 females worldwide. Because males die sooner, though, the adult ratio of men to women is about one to one.

There is terrible inequality between the two chromosomes. The X chromosome carries about 1,500 genes, which do most of the heavy lifting to develop an embryo. The little Y chromosome, by comparison, has been shedding its associated genes at a rate of about five every one million years. It's now down to less than 100 genes.

With only a single X chromosome, males need every one of those 1,500 genes. With two X chromosomes, females have double the necessary amount. You can think of it like a cake recipe calling for only one cup of flour. If you decide to put in two, it will change the results in a most unpleasant fashion. The female embryo uses what may be the most time-honored weapon in the battle of the sexes to solve the problem of two Xs: She simply ignores one of them. This chromosomal silent treatment is known as X inactivation. One of the chromosomes is tagged with the molecular equivalent of a "Do Not Disturb" sign. Because males require all 1,500 X genes to survive, and they have only one X chromosome, X inactivation does not occur in guys. And because males must get their X from Mom, all men are literally, with respect to their X chromosome, Momma's Boys—unisexed.

That's very different from their sisters, who are more genetically complex. Since female embryos have two Xs from which to choose, Mom's or Dad's, researchers wanted to know who preferentially got the sign. The answer was completely unexpected: *There were no preferences.* Some cells in the developing little girl embryo hung their sign around Mom's X. Neighboring cells hung their sign around Dad's. At this point in research, there doesn't appear to be any rhyme or reason, and it is considered a random event. This means that cells in the female embryo are a complex mosaic of both active and inactive mom-and-pop X genes. These bombshells describe our first truly genetic-based findings of potential differences between men's and women's brains.

What do many of the X's 1,500 genes do? They govern how we think. In 2005 the human genome was sequenced, and a large percentage of the X chromosome genes were found to create proteins involved in brain manufacture. Some of these genes may be involved in establishing higher cognitive functions, from verbal skills and social behavior to certain types of intelligence. Researchers call the X chromosome a cognitive "hot spot."

The purpose of genes is to create molecules that mediate the functions of the cells in which they reside. Collections of these cells create the large brain structures we've been talking about, like the cortex, the hippocampus, the thalamus and the amygdala. These make up the neuroanatomy of the brain, which we turn to next.

Differences in brain structure

When it comes to neuroanatomy, the real challenge is finding areas that *aren't* affected by sex chromosomes. You can see differences in the cortex, the amygdala, even the biochemicals that brain cells use to communicate with each other.

The frontal and prefrontal cortex control much of our decision-making ability. Labs—headed by scientists of both sexes, I should perhaps point out—have found that certain parts of this cortex is fatter in women than in men.

The limbic system, home to the amygdala, controls not only the generation of emotions but also the ability to remember them. Running counter to current social prejudice, this region is much larger in men than it is in women. At rest, female amygdalas tend to talk mostly to the left hemisphere, while male amygdalas do most of their chatting with the right hemisphere.

Biochemicals have not escaped sex differences, either. Serotonin, key in regulating emotion and mood, is a particularly dramatic example. Males can synthesize serotonin about 52 percent faster than females. (Prozac works by altering the regulation of this neurotransmitter.)

What do these physical differences really mean? In animals, the size of structures is thought to reflect their relative importance to survival. Human examples at first blush seem to follow a similar pattern. We already have noticed that violinists have bigger areas of the brain devoted to controlling their left hand than their right. But neuroscientists nearly come to blows over how brain structure relates to function. We don't yet know whether differences in the size of a brain region translate to anything substantial when it comes to behavior.

Differences in behavior

I didn't really want to write about this. Characterizing gender-specific behaviors has a long and mostly troubled history. Institutions holding our best minds aren't immune. Larry Summers was *Harvard's* president, for Pete's sake, when in 2005 he attributed girls' lower math and science scores to behavioral genetics, comments that cost him his job. The battle of the sexes has existed for a very long time, illustrated by three quotes separated by centuries:

"The female is an impotent male, incapable of making semen because of the coldness of her nature. We therefore should look upon the female state as if it were a deformity, though one that occurs in the ordinary course of nature."

Aristotle (384–332 BC)

"Girls begin to talk and to stand on their feet sooner than boys because weeds always grow up more quickly than good crops."

Martin Luther (1483–1546)

"If they can put a man on the moon ... why can't they put them all there?"
Jill (graffiti I saw on a bathroom wall in 1985, in response to Luther's quote scribbled there)

Almost 2,400 years of history separate Aristotle from Jill, yet we seem to have barely moved. Invoking planet metaphors like

Venus and Mars, some purport to expand perceived differences into prescriptions for relationships. And this is the most scientifically progressive era in human history.

Mostly, I think, it comes down to statistics. When people hear about measurable differences, they often think scientists are talking about individuals, such as themselves. That's a *big* mistake. When scientists look for behavioral trends, they do not look at individuals. They look at populations. Trends emerge, but the many variations and overlaps mean that statistics in these studies can never apply to individuals. There may very well be differences in the way men and women think about some things. But exactly how that relates to your behavior is a completely separate question.

Mental disorders

Brain pathologies represent one of the strongest pieces of evidence that sex chromosomes are involved in brain function and thus brain behavior. Mental retardation is more common in males than in females in the general population. Many of these pathologies are caused by mutations in any one of 24 genes in the X chromosome. As you know, males have no backup X. If their X gets damaged, they have to live with the consequences. If a female's X is damaged, she can often ignore the consequences.

Mental-health professionals have known for years about sex-based differences in the type and severity of psychiatric disorders. Males are more severely afflicted by schizophrenia than females, for example. By more than two to one, women are more likely to get depressed than men, a figure that shows up just after puberty and remains stable for the next 50 years. Males have a greater tendency to be antisocial. Females have more anxiety. Most alcoholics and drug addicts are male. Most anorexics are female. Says Thomas Insel, from the National Institute of Mental Health, "It's pretty difficult to find any single factor that's more predictive for some of these disorders than gender."

Emotions and stress

It's a horrible slide show. In it, a little boy is run over by a car while walking with his parents. If you ever see that show, you will never forget it. But what if you *could* forget it? The brain's amygdala aids in the creation of emotions and our ability to remember them. Suppose there was a magic elixir that could momentarily suppress it? Such an elixir does exist, and it was used to show that men and women process emotions differently.

You have probably heard the term left brain versus right brain. You may have heard that this underscores creative versus analytical people. That's a folk tale, the equivalent of saying the left side of a luxury liner is responsible for keeping the ship afloat, and the right is responsible for making it move through the water. Both sides are involved in both processes. That doesn't mean the hemispheres are equal, however. The right side of the brain tends to remember the gist of an experience, and the left brain tends to remember the details.

Researcher Larry Cahill eavesdropped on men's and women's brains under acute stress (he showed them slasher films), and what he found is this: Men handled the experience by firing up the amygdala in their brain's right hemisphere. Their left was comparatively silent. Women handled the experience with the opposite hemisphere. Their left amygdala lit up, their right comparatively silent. If males are firing up the side in charge of gist, does that mean males remember more gist than detail of a given emotional experience related to stress? Conversely, do females remember more detail than gist? Cahill decided to find out.

That magic elixir of forgetting, a drug called propranolol, normally is used to regulate blood pressure. As a beta-blocker, it also inhibits the biochemistry that activates the amygdala during emotional experiences. The drug is being investigated as a potential treatment for combat-related disorders.

But Cahill gave it to his subjects before they watched a traumatic film. One week later, he tested their memories of it. Sure enough, the men lost the ability to recall the gist of the story, compared with men who didn't take the drug. Women lost the ability to recall the details. One must be careful not to overinterpret these data. The results clearly define only emotional responses to stressful situations, not objective details and summaries. This is not a battle between the accountants and the visionaries.

Cahill's results come on the heels of similar findings around the world. Other labs have extended his work, finding that women recall more emotional autobiographical events, more rapidly and with greater intensity, than men do. Women consistently report more vivid memories for emotionally important events such as a recent argument, a first date, or a vacation. Other studies show that, under stress, women tend to focus on nurturing their offspring, while men tend to withdraw. This tendency in females has sometimes been called "tend and befriend." Is this caused by nature or nuture? As Stephen Jay Gould says, "It is logically, mathematically, and philosophically impossible to pull them apart."

Verbal communication

Over the past several decades, behaviorist Deborah Tannen and others have done some fascinating work on how men and women communicate verbally. The CliffsNotes version of their findings: Women are better at it.

Women tend to use both hemispheres when speaking and processing verbal information. Men primarily use one. Women tend to have thick cables connecting their two hemispheres. Men's are thinner. It's as though females have a backup system that males don't. Researchers think these neuroanatomical differences may explain why language and reading disorders occur approximately twice as often in little boys as in little girls. Women also recover from stroke-induced verbal impairment better than men.

Girls seem verbally more sophisticated than little boys as they go through the school system. They are better at verbal memory tasks, verbal fluency tasks, and speed of articulation. When these little girls grow up, they are still champions at processing verbal information. Real as these data seem, however, almost none of them can be divorced from a social context. That's why Gould's comment is so helpful.

Tannen spent years observing and videotaping how little girls and little boys interact, especially when talking to their best friends. If any detectable patterns emerged in children, she wanted to know if they also showed up in college students. The patterns she found were both predictable and stable. The conversational styles we develop as adults come directly from the same-sex interactions we solidified as children. Tannen's findings center on how boys and girls cement relationships and negotiate status within same-sex groups, and then how these entrenched styles clash as men and women try to communicate with one another as adults.

Cementing relationships

When girl best friends communicate with each other, they lean in, maintain eye contact, and do a lot of talking. They use their sophisticated verbal talents to cement their relationships. Boys never do this. They rarely face each other directly, preferring either parallel or oblique angles. They make little eye contact, their gaze always casting about the room. They do not use verbal information to cement their relationships. Instead, commotion seems to be the central currency of a little boy's social economy. Doing things physically together is the glue that cements their relationships.

My sons, Josh and Noah, have been playing a one-upmanship game since they were toddlers. A typical version might involve ball throwing. Josh would say, "I can throw this up to the ceiling," and would promptly try. Then they would laugh. Noah would respond by grabbing and throwing the ball, saying, "Oh yeah? I can throw this up to the sky." This ratcheting, with laughter, would continue until they reached the "galaxy" or the big prize, "God."

Tannen consistently saw this style everywhere she looked—except when observing little girls. The female version goes something like this. One sister says, "I can take this ball and throw it to the ceiling," and she does. Both laugh. The other sister grabs the ball, throws it up to the ceiling, and says, "I can, too!" Then they talk about how cool it is that they can both throw the ball. This style persists into adulthood for both sexes. Tannen's data, unfortunately, have been misinterpreted as "Boys always compete, and girls always cooperate." As this example shows, however, boys are being extremely cooperative. They are simply doing it through competition, deploying their favorite strategy of physical activity.

Negotiating status

By elementary school, boys finally start using their verbal skills for something: to negotiate their status in a large group. Tannen found that high-status males give orders to the rest of the group, verbally or even physically pushing the low-status boys around. The "leaders" maintain their fiefdoms not only by issuing orders but also by making sure the orders are carried out. Other strong members try to challenge them, so the guys at the top learn quickly to deflect challenges. Hierarchy is very evident with boys. It can be hard on them, too: The life of a low-status male is often miserable.

Tannen found that little girls have hierarchies of status, too. But they used strikingly different strategies to generate and maintain them. Verbal communication is so important that the type of talk determines the status of the relationship. Your "best friend" is the one to whom you tell secrets. The more secrets revealed, the more likely the girls are to identify each other as close. Girls tend to deemphasize the status between them in these situations. Using their sophisticated verbal ability, the girls tend not to give top-down imperial orders. If one of the girls tries issuing commands, the style is usually rejected: The girl is tagged as "bossy" and isolated socially. Not that decisions aren't made. Various members of the group give suggestions, discuss alternatives, and come to a consensus.

The difference between girls' and boys' communication could be described as the addition of a single powerful word. Boys might say, "Do this." Girls would say, "*Let's* do this."

Styles persist, then clash

Tannen found that, over time, these ways of using language become increasingly reinforced. By college age, most of these styles are deeply entrenched. And that's when the problems between men and women become most noticeable.

Tannen tells the story of a woman driving with her husband. "Would you like to stop for a drink?" the wife asked. The husband wasn't thirsty. "No," he replied. The woman was annoyed because she had wanted to stop; the man was annoyed because she wasn't direct. In her book *You Just Don't Understand*, Tannen explains: "From her point of view, she had shown concern for her husband's wishes, but he had shown no concern for hers." How would this conversation likely go between two women? The thirsty woman would ask, "Are you thirsty?" With lifelong experience at verbal negotiation, her friend would know what she wanted and respond, "I don't know. Are *you* thirsty?" Then a small discussion would ensue about whether they were both thirsty enough to stop the car and get water.

These differences in social sensitivity play out in the workforce just as easily as in marriage. At work, women who exert "male" leadership styles are in danger of being perceived as bossy and aggressive. Men who do the same thing are often praised as decisive and assertive. Tannen's great contribution was to show that these stereotypes form very early in our social development, perhaps assisted by asymmetric verbal development. They transcend geography, age, and even time. Tannen, who was an English literature major, sees these tendencies in manuscripts that go back centuries.

Nature or nurture?

Tannen's findings are statistical patterns, not an all-or-none phenomenon. Many factors affect our language patterns, she found. Regional background, individual personality, profession, social class, age, ethnicity, and birth order all affect how we use language to negotiate our social ecologies. Boys and girls are treated differently socially the moment they are born, and they are often reared in societies filled with centuries of entrenched prejudice. It would be a miracle if we somehow transcended our experience and behaved in an egalitarian fashion.

Given the influence of culture on behavior, it is overly simplistic to invoke a purely biological explanation for Tannen's observations. Given the great influence of brain biology on behavior, it is also simplistic to invoke a purely social explanation. The real answer to the nature-or-nurture question is "We don't know." That can be frustrating to hear. As scientists explore how genes and cells and behaviors connect, their findings give us not completed bridges but boards and nails. It's dangerous to assume the bridges are complete. Just ask Larry Summers.

More ideas

Get the facts straight on emotions

Dealing with the emotional lives of men and women is a big part of the job for teachers and managers. They need to know:

1) Emotions are useful. They make the brain pay attention.

2) Men and women process certain emotions differently.

3) The differences are a product of complex interactions between nature and nurture.

Experiment with same-sex classrooms

My son's third-grade teacher began seeing a stereotype that worsened as the year progressed. The girls were excelling in the language arts, and the boys were pulling ahead in math and science. This was only the *third grade!* The language-arts differences made some sense to her. But she knew there was no statistical support for the contention that men have a better aptitude for math and science than women. Why, for heaven's sake, was she presiding over a stereotype? The teacher guessed that part of the answer lay in the students' social participation during class. When the teacher asked a question of the class, who answered first turned out to be unbelievably important. In the language arts, the girls invariably answered first. Other girls reacted with that participatory "me too" instinct. The reaction on the part of the boys was hierarchical. The girls usually knew the answers, the boys usually did not, and the males responded by doing what low-status males tend to do: They withdrew. A performance gap quickly emerged.

In math and science, boys and girls were equally likely to answer a question first. But the boys used their familiar "top each other" conversational styles when they participated, attempting to establish a hierarchy based on who knew more. This included drubbing anyone who didn't make the top, including the girls. Bewildered, the girls began withdrawing from participating in the subjects. Once again, a performance gap emerged.

The teacher called a meeting of the girls and verified her observations. Then she asked for a consensus about what they should do. The girls decided that they wanted to learn math and science separately from the boys. Previously a strong advocate for mixed-sex classes, the teacher wondered aloud if that made any sense. Yet if the girls started losing the math-and-science battle in the third grade, the teacher reasoned, they were not likely to excel in the coming years. She obliged. It took only two weeks to close the performance gap.

Can the teacher's result be applied to classrooms all over the world? One classroom in a single school year does not make for a valid experiment. We need to test hundreds of classrooms and thousands of students from all walks of life, over a period of years.

Pair men and women in workplace teams

One day, I spoke about gender with a group of executives-in-training at the Boeing Leadership Center in St. Louis. After showing some of Larry Cahill's data about gist and detail, I said, "Sometimes women are accused of being more emotional than men, from the home to the workplace. I think that women might not be any more emotional than anyone else." I explained that because women perceive their emotional landscape with more data points (that's the detail) and see it in greater resolution, women may simply have more information to which they are capable of reacting. If men perceived the same number of data points, they might have the same reactions. Two women in the back began crying softly. After the lecture, I asked them about it, fearing I may have offended them. What they said instead blew me away. "It was the first time in my professional life," one of them said, "that I didn't feel like I had to apologize for who I was."

And that got me to thinking. In our evolutionary history, having a team that could understand both the gist and details of a given stressful situation helped us conquer the world. Why would the world of business be exempted from that advantage? Having an executive team or work group capable of simultaneously understanding both the emotional forests and the trees of a stressful project, such as a merger, might be a marriage made in business heaven. It could even affect the bottom line.

Companies often train managers by setting up simulations of various situations. They could take a mixed-sex team and a unisex team and have each work on the same project. Give another two teams the same project, but first teach them what we know about brain differences between the sexes. Would the mixed teams do better than the unisex teams? Would the groups prepped on how the brain works do better than the unprepped groups? You might find that management teams with a gist/detail balance create the best chance for productivity. At the very least, it means both men and women have an equal right to be at the decision-making table.

Imagine environments where gender differences are both noted and celebrated, as opposed to ignored and marginalized. We might have more women in science and engineering. We might shatter the archetypal glass ceiling. We might create better businesses. We might even create better marriages.

Brain Rule #11
Male and female brains are different.

* The X chromosome that males have one of and females have two of—though one acts as a backup—is a cognitive "hot spot," carrying an unusually large percentage of genes involved in brain manufacture.

* Women are genetically more complex, because the active X chromosomes in their cells are a mix of Mom's and Dad's. Men's X chromosomes all come from Mom, and their Y chromosome carries less than 100 genes, compared with about 1,500 for the X chromosome.

* Men's and women's brains are different structurally and biochemically—men have a bigger amygdala and produce serotonin faster, for example—but we don't know if those differences have significance.

* Men and women respond differently to acute stress: Women activate the left hemisphere's amygdala and remember the emotional details. Men activate the right hemisphere's amygdala and get the gist.

exploration

Brain Rule #12

Brain Rule #12
We are powerful and natural explorers.

MY DEAR SON JOSH got a painful beesting at the tender age of 2, and he almost deserved it.

It was a warm, sunny afternoon. We were playing the "pointing game," a simple exercise where he would point at something, and I would look. Then we'd both laugh. Josh had been told not to touch bumblebees because they could sting him; we used the word "danger" whenever he approached one. There, in a patch of clover, he spotted a big, furry, buzzing temptress. As he reached for it, I calmly said, "Danger," and he obediently withdrew his hand. He pointed at a distant bush, continuing our game.

As I looked toward the bush, I suddenly heard a 110-decibel yelp. While I was looking away, Josh reached for the bee, which promptly stung him. Josh had used the pointing game as a diversion, and I was outwitted by a 2-year-old.

"DANGER!" he sobbed as I held him close.

"Danger," I repeated sadly, hugging him, getting some ice, and wondering what puberty would be like in 10 years or so.

This incident was Dad's inauguration into a behavioral suite often called the terrible twos. It was a rough baptism for me and the little guy. Yet it also made me smile. The mental faculties kids use to distract their dads are the same they will use as grown-ups to discover the composition of distant suns or the next alternative energy. We are natural explorers, even if the habit sometimes stings us. The tendency is so strong, it is capable of turning us into lifelong learners. But you can see it best in our youngest citizens, often when they seem at their worst.

Babies give researchers a clear view, unobstructed by years of contaminating experiences, of how humans naturally acquire information. Preloaded with lots of information-processing software, infants acquire information using surprisingly specific strategies, many of which are preserved into adulthood. In part, understanding how humans learn at this age means understanding how humans learn at any age.

We didn't always think that way. If you had said something about preset brain wiring to researchers 40 years ago, their response would have been an indignant, "What are you smoking?" or, less politely, "Get out of my laboratory." This is because researchers for decades thought that babies were a blank slate—a tabula rasa. They thought that everything a baby knew was learned by interactions with its environments, primarily with adults. This perspective undoubtedly was formulated by overworked scientists who never had any children. We know better now. Amazing strides have been made in understanding the cognitive world of the infant. Indeed, the research world now looks to babies to show how humans, including adults, think about practically everything.

Babies test everything—including you

Babies are born with a deep desire to understand the world around them and an incessant curiosity that compels them to aggressively explore it. This need for explanation is so powerfully stitched into their experience that some scientists describe it as a drive, just as hunger and thirst and sex are drives.

All babies gather information by actively testing their environment, much as a scientist would. They make a sensory observation, form a hypothesis about what is going on, design an experiment capable of testing the hypothesis, and then draw conclusions from the findings. They use a series of increasingly self-corrected ideas to figure out how the world works.

42 minutes old: Newborns can imitate

In 1979, Andy Meltzoff rocked the world of infant psychology by sticking out his tongue at a newborn and being polite enough to wait for a reply. What he found astonished him. The baby stuck her tongue back out at him! He reliably measured this imitative behavior with infants only 42 minutes old. The baby had never seen a tongue before, not Meltzoff's and not her own, yet the baby knew she had a tongue, knew Meltzoff had a tongue, and somehow intuited the idea of mirroring. Further, the baby's brain knew that if it stimulated a series of nerves in a certain sequence, she could also stick her tongue out. That's definitely not consistent with the notion of tabula rasa.

I tried this with my son Noah. He and I started our relationship in life by sticking our tongues out at each other. In his first 30 minutes of life, we had struck up an imitative conversation. By the end of his first week, we were well entrenched in dialogue: Every time I came into his crib room, we greeted each other with tongue protrusions. It was purely delightful on my part and purely adaptive on his. If I had not stuck my tongue out initially, he would not be doing so with such predictability every time he saw me.

Three months later, my wife picked me up after a lecture at a medical school, Noah in tow. I was still fielding questions, but I scooped up Noah and held him close while answering. Out of the corner of my eye, I noticed Noah gazing at me expectantly, flicking his tongue out about every five seconds. I smiled and stuck my tongue out at Noah mid-question. Immediately he squealed and started sticking his tongue out with abandon, every half second or so. I knew exactly what he was doing. Noah made an observation (Dad and I stick our tongues out at each other), formed a hypothesis (I bet if I stick my tongue out at Dad, he will stick his tongue back out at me), created and executed his experiment (I will stick my tongue out at Dad), and changed his behavior as a result of the evaluation of his research (sticking his tongue out more frequently).

Nobody taught Noah, or any other baby, how to do this. And it is a lifelong strategy. You probably did it this morning when you couldn't find your glasses, hypothesized they were in the bathroom, and went to look. From a brain science perspective, we don't even have a good metaphor to describe how you know to do that. It is so automatic, you probably had no idea you were looking at the results of a successful experiment when you found your glasses lying on a towel.

Noah's story is just one example of how babies use their precious preloaded information-gathering strategies to gain knowledge they didn't have at birth. We also can see it in broken stuff, disappearing cups and temper tantrums.

12 months old: Infants analyze how objects act

Babies younger than a year old will systematically analyze an object with every sensory weapon at their disposal. They will feel it, kick it, try to tear it apart, stick it in their ear, stick it in their mouth, give it to you so that you can stick it in your mouth. They appear to be intensely gathering information about the properties of the object. Babies methodically do experiments on the objects to see what else they will do. In our household, this usually meant breaking stuff.

These object-oriented research projects grow increasingly sophisticated. In one set of experiments, babies were given a rake, and a toy was placed nearby. The babies quickly learned to use the rake to get the toy. This is not exactly a groundbreaking discovery, as every parent knows. After a few successful attempts, the babies lost interest in the toy. But not in the experiment. Again and again, they would take the toy and move it to a different place, then use the rake to grab it. You can almost hear them exclaiming, "Wow! How does this happen?"

18 months old: Objects still exist if you can't see them

Little Emily, before 18 months of age, still believes that if an object is hidden from view, that object has disappeared. She does not have what is known as "object permanence." That is about to change. Emily has been playing with a washcloth and a cup. She covers the cup with the cloth, and then pauses for a second, a concerned look on her brow. Slowly she pulls the cloth away from the cup. The cup is still there! She glares for a moment, then quickly covers it back up. Thirty seconds go by before her hand tentatively reaches for the cloth. Repeating the experiment, she slowly removes the cloth. The cup is *still* there! She squeals with delight. Now things go quickly. She covers and uncovers the cup again and again, laughing loudly each time. It is dawning on Emily that the cup has object permanence: Even if removed from view, it has not disappeared. She will repeat this experiment for more than half an hour. If you have ever spent time with an 18-month-old, you know that getting one to concentrate on anything for 30 minutes is some kind of miracle. Yet it happens, and to babies at this age all over the world.

Though this may sound like a delightful form of peekaboo, it is actually an experiment whose failure would have lethal evolutionary consequences. Object permanence is an important concept to have if you live in the savannah. Saber-toothed tigers still exist, for example, even if they suddenly duck down in the tall grass. Those who didn't acquire this knowledge usually were on some predator's menu.

18 months old: Your preferences aren't the same as mine

The distance between 14 months of age and 18 months of age is extraordinary. Around 14 months, toddlers think that because they like something, the whole world likes the same thing—as summed up in the "Toddler's Creed":

> *If I want it, it is mine.*
> *If I give it to you and change my mind later, it is mine.*
> *If I can take it away from you, it is mine.*
> *If we are building something together, all of the pieces are mine.*
> *If it looks just like mine, it is mine.*
> *If it is mine, it will never belong to anybody else, no matter what.*
> *If it is yours, it is mine.*

Around 18 months, it dawns on babies that this viewpoint may not always be accurate. They begin to learn that adage that most newlyweds have to relearn in spades: "What is obvious to you is obvious to you."

How do babies react to such new information? By testing it, as usual. Before the age of 2, babies do plenty of things parents would rather them not do. But after the age of 2, small children will do things *because* their parents don't want them to. The compliant little darlings seem to transform into rebellious little tyrants. Many parents think their children are actively defying them at this stage. (The thought certainly crossed my mind as I nursed Joshua's unfortunate beesting.) That would be a mistake, however. This stage is simply the natural extension of a sophisticated research program begun at birth. You push the boundaries of people's preferences, then stand back and see how they react. Then you repeat the experiment, pushing them to their limits over and over again to see how stable the findings are, as if you were playing peekaboo. Slowly you begin to perceive the length and height and breadth of people's desires, and how they differ from yours. Then, just to be sure the boundaries are still in place, you occasionally do the whole experiment over again.

Babies may not have a whole lot of understanding about their world, but they know a whole lot about how to get it. It reminds me of the old proverb, "Catch me a fish and I eat for a day; teach me to fish and I eat for a lifetime."

Babies reveal more of the brain's secrets each year

Why does a baby stick its tongue back out at you? The beginnings of a neural road map have been drawn in the past few years, at least for some of the "simpler" thinking behaviors, such as imitation. Three investigators at the University of Parma were studying the macaque, assessing brain activity as it reached for different objects in the laboratory. The researchers recorded the pattern of neural firing when the monkey picked up a raisin. One day, researcher Leonardo Fogassi walked into the laboratory and casually plucked a raisin from a bowl. Suddenly, the monkey's brain began to fire excitedly. The recordings were in the raisin-specific pattern, *as if the animal had just picked up the raisin.* But the monkey had not picked up the raisin. It simply saw Fogassi do it.

The astonished researchers quickly replicated and extended their findings, and then published them in a series of landmark papers describing the existence of "mirror neurons." Mirror neurons are cells whose activity reflect their surroundings. Cues that could elicit mirror neural responses were found to be remarkably subtle. If a primate simply heard the sound of someone doing something it had previously experienced—say, tearing a piece of paper—these neurons could fire as if the monkey were experiencing the full stimulus. It wasn't long before researchers identified human mirror neurons. These neurons are scattered across the brain, and a subset is involved in action recognition—that classic imitative behavior such as babies sticking out their tongues. Other neurons mirror a variety of motor behaviors.

251

We also are beginning to understand which regions of the brain are involved in our ability to learn from a series of increasingly self-corrected ideas. We use our right prefrontal cortex to predict error and to retrospectively evaluate input for errors. The anterior cingulate cortex, just south of the prefrontal cortex, signals us when perceived unfavorable circumstances call for a change in behavior. Every year, the brain reveals more and more of its secrets, with babies leading the way.

We never outgrow the desire to know

We can remain lifelong learners. No question. This fact was brought home to me as a postdoctoral scholar at the University of Washington. In 1992, Edmond Fischer and Edwin Krebs shared the Nobel Prize in Physiology or Medicine. I had the good fortune to be familiar with both their work and their offices. They were just down the hall from mine. By the time I arrived at the university, Fischer and Krebs were already in their mid-70s. The first thing I noticed upon meeting them was that they were not retired. Not physically and not mentally. Long after they had earned the right to be lounging on some tropical island, both had powerful, productive laboratories in full swing. Every day I would see them walking down the hall, oblivious to others, chatting about some new finding, swapping each other's journals, and listening intently to each other's ideas. Sometimes they would have someone else along, grilling them and in turn being grilled about some experimental result. They were as creative as artists, wise as Solomon, lively as children. They had lost *nothing*. Their intellectual engines were still revving, and curiosity remained the fuel. They taught me that our learning abilities don't have to change as we age.

The brain remains malleable

Research shows that the brain is wired to keep learning as we age. Some regions of the adult brain stay as malleable as a baby's brain, so we can grow new connections, strengthen existing connections, and even create new neurons, allowing all of us to be lifelong learners. We didn't always think that. Until five or six years ago, the prevailing notion was that we were born with all of the brain cells we were ever going to get, and they steadily eroded in a depressing journey through adulthood to old age. We do lose synaptic connections with age. Some estimates of neural loss alone are close to 30,000 neurons per day. But the adult brain also continues creating neurons within the regions normally involved in learning. These new neurons show the same plasticity as those of newborns.

Throughout life, your brain retains the ability to change its structure and function in response to your experiences.

Why? Evolutionary pressure, as usual. Problem solving was greatly favored in the unstable environment of the Serengeti. But not just any kind of problem solving. When we came down from the trees to the savannah, we did not say to ourselves, "Good Lord, give me a book and a lecture and a board of directors so that I can spend 10 years learning how to survive in this place." Our survival did not depend upon exposure to organized, preplanned packets of information. Our survival depended upon chaotic, reactive information-gathering experiences. That's why one of our best attributes is the ability to learn through a series of increasingly self-corrected ideas. "The red snake with the white stripe bit me yesterday, and I almost died," is an observation we readily made. Then we went a step further: "I hypothesize that if I encounter the same snake, the same thing will happen!" It is a scientific learning style we have exploited literally for millions of years. It is not possible to outgrow it in the whisper-short seven to eight decades we spend on the planet.

So it's possible for us to continue exploring our world as we age. Of course, we don't always find ourselves in environments that encourage such curiosity as we grow older. I've been fortunate to have a career that allowed me the freedom to pick my own projects. Before that, I was lucky to have my mother.

Encouraging curiosity with a passion

I remember, when I was 3 years old, obtaining a sudden interest in dinosaurs. I had no idea that my mother had been waiting for it. That very day, the house began its transformation into all things Jurassic. And Triassic. And Cretaceous. Pictures of dinosaurs would go up on the wall. I would begin to find books about dinosaurs strewn on the floor and sofas. Mom would even call dinner "dinosaur food," and we would spend hours laughing our heads off trying to make dinosaur sounds. And then, suddenly, I would lose interest in dinosaurs, because some friend at school acquired an interest in spaceships and rockets and galaxies. Extraordinarily, my mother was waiting. Just as quickly as my whim changed, the house would begin its transformation from big dinosaurs to Big Bang. The reptilian posters came down, and in their places, planets would begin to hang from the walls. I would find little pictures of satellites in the bathroom. Mom even got "space coins" from bags of potato chips, and I eventually gathered all of them into a collector's book.

This happened over and over again in my childhood. I got an interest in Greek mythology, and she transformed the house into Mount Olympus. My interests careened into geometry, and the house became Euclidean, then cubist. Rocks, airplanes. By the time I was 8 or 9, I was creating my own house transformations.

One day, around age 14, I declared to my mother that I was an atheist. She was a devoutly religious person, and I thought this announcement would crush her. Instead, she said something like "That's nice, dear," as if I had just declared I no longer liked nachos. The next day, she sat me down by the kitchen table, a wrapped package in her lap. She said calmly, "So I hear you are now an atheist. Is that true?" I nodded yes, and she smiled. She placed the package in my hands. "The man's name is Friedrich Nietzsche, and the book is called *Twilight of the Idols*," she said. "If you are going to be an atheist, be the best one out there. *Bon appetit!*"

I was stunned. But I understood a powerful message: Curiosity itself was the most important thing. And what I was interested in *mattered*. I have never been able to turn off this fire hose of curiosity.

Most developmental psychologists believe that a child's need to know is a drive as pure as a diamond and as distracting as chocolate. Even though there is no agreed-upon definition of curiosity in cognitive neuroscience, I couldn't agree more. I firmly believe that if children are allowed to remain curious, they will continue to deploy their natural tendencies to discover and explore until they are 101. This is something my mother seemed to know instinctively.

For little ones, discovery brings joy. Like an addictive drug, exploration creates the need for more discovery so that more joy can be experienced. It is a straight-up reward system that, if allowed to flourish, will continue into the school years. As children get older, they find that learning brings them not only joy but also mastery. Expertise in specific subjects breeds the confidence to take intellectual risks. If these kids don't end up in the emergency room, they may end up with a Nobel Prize.

I believe it is possible to break this cycle, anesthetizing both the process and the child. By first grade, for example, children learn that education means an A. They begin to understand that they can acquire knowledge not because it is interesting, but because it can get them something. Fascination can become secondary to "What do I need to know to get the grade?" But I also believe the curiosity instinct is so powerful that some people overcome society's message to go to sleep intellectually, and they flourish anyway.

My grandfather was one of those people. He was born in 1892 and lived to be 101 years old. He spoke eight languages, went through several fortunes, and remained in his own house (mowing his own lawn) until the age of 100. He was lively as a firecracker to the end. At a party celebrating his centenary, he took me aside. "You know, *Juanito*," he said, clearing his throat, "sixty-six years separate the Wright brothers' airplane from Neil Armstrong and the moon." He shook his head, marveling. "I was born with the horse and buggy. I die with the space shuttle. What kind of thing is that?" His eyes twinkled. "I live the good life!"

He died a year later.

I think of him a lot when I think of exploration. I think of my mother and her magically transforming rooms. I think of my youngest son experimenting with his tongue, and my oldest son's overwhelming urge to take on a beesting. And I think that we must do a better job of encouraging lifelong curiosity, in our workplaces, our homes, and especially in our schools.

More ideas

On a personal level, what this tells us is to follow our passions. But I would also like to see change on a broader scale so that our environments truly support our individual efforts to remain curious.

Free time at work

Smart companies take to heart the power of exploration. For example, companies such as 3M, Genentech, and Google allowed employees to use 15 or 20 percent of their workweek to go where their mind asks them to go. The proof is in the bottom line: At Google, fully 50 percent of new products—including Gmail, Google News, and AdSense—came from "20 percent time." Facebook, LinkedIn, and other tech companies hold "hackathons": marathon programming sessions where coders can earn prizes for creating something interesting.

Schools where you learn on the job

If you could step back in time to one of the first Western-style universities, say, the University of Bologna, and visit its biology labs, you would laugh out loud. I would join you. By today's standards, biological science in the 11th century was a joke, a mix of astrological influences, religious forces, dead animals, and rude-smelling chemical concoctions. But if you went down the hall and peered inside Bologna's standard lecture room, you wouldn't feel as if you were in a museum. You would feel at home. There is a lectern for the teacher to hold forth, surrounded by chairs for the students to absorb whatever is being held forth—much like today's classrooms. Could it be time for a change?

Some people have tried to harness our natural exploratory tendencies by using "problem-based" or "discovery-based" learning models. What's missing are empirical results that show the long-term effects of these styles. To this end, I would like to see more degree programs modeled after medical schools. The best medical-school model has three components: a teaching hospital; faculty who work in the field as well as teach; and research laboratories. It is a surprisingly successful way of transferring complex information from one brain to another. Students get consistent exposure to the real world, by the third year spending half of their time in class and half learning on the job. They are taught by people who actually do what they teach as their "day job." And they get to participate in practical research programs.

Here's a typical experience in medical school: The clinician-professor is lecturing in a traditional classroom setting and brings in a patient to illustrate some of his points. The professor announces: "Here is the patient. Notice that he has disease X with symptoms A, B, C, and D." He then begins to lecture on the biology of disease X. While everybody is taking notes, a smart medical student raises her hand and says, "I see symptoms A, B, C, and D. What about symptoms E, F, and G?" The professor looks a bit chagrined (or excited) and responds, "We don't know about symptoms E, F, and G." You can hear a pin drop at those moments, and the impatient voices whispering inside the students' heads are almost audible: "Well, let's find out!" These are the opening words of most of the great research ideas in human medicine.

That's true exploratory magic. The tendency is so strong that you have to deliberately cut off the discussions to keep the ideas from forming. Rather than cutting off such discussions, most American medical schools possess powerful research wings. By simple juxtaposition of real-world needs with traditional book learning, a research program is born.

I envision a college of education where the program is all about brain development. Like a medical school, it is divided into three parts. It has traditional classrooms. It is a community school staffed and run by three types of faculty: traditional education faculty who teach the college students, certified teachers who teach the little ones attending the community school, and brain scientists who run the research labs devoted to a single purpose: investigating how the human brain learns in teaching environments, then actively testing hypothesized ideas in real-world classroom situations.

Students would get a bachelor of *science* in education. Future educators are infused with deep knowledge about how the human brain acquires information. After their first year of study, students would start actively participating in the on-site school.

This model honors our evolutionary need to explore. It creates teachers who know about brain development. And it's a place to do the real-world research so sorely needed to figure out how, exactly, the rules of the brain should be applied to our lives. The model could apply to other academic subjects as well. A business school teaching how to run a small business might actually run one, for example.

A student could create a version of this learning experience on her own, by seeking out internship opportunities while in school.

A sense of wonder

My 2-year-old son Noah and I were walking down the street on our way to preschool when he suddenly noticed a shiny pebble embedded in the concrete. Stopping midstride, the little guy considered it for a second, found it thoroughly delightful, and let out a laugh. He spied a small plant an inch farther, a weed valiantly struggling through a crack in the asphalt. He touched it gently, then laughed again. Noah noticed beyond it a platoon of ants marching in single file, which he bent down to examine closely. They were carrying a dead bug, and Noah clapped his hands in wonder. There were dust particles, a rusted screw, a shiny spot of oil. Fifteen minutes had passed, and we had gone only 20 feet. I tried to get him to move along, having the audacity to act like an adult with a schedule. He was having none of it. And I stopped, watching my little teacher, wondering how long it had been since I had taken 15 minutes to walk 20 feet.

The greatest Brain Rule of all is something I cannot prove or characterize, but I believe in it with all my heart. As my son was trying to tell me, it is the importance of curiosity. For his sake and ours, I wish classrooms and companies were designed with the brain in mind. If we started over, curiosity would be the most vital part of both demolition crew and reconstruction crew. As I hope to have related here, I am very much in favor of both.

I will never forget the moment my little professor taught his daddy about what it meant to be a student. I was thankful and a little embarrassed. After 47 years, I was finally learning how to walk down the street.

Brain Rule #12
We are powerful and natural explorers.

- Babies are the model of how we learn—not by passive reaction to the environment but by active testing through observation, hypothesis, experiment, and conclusion.

- Specific parts of the brain allow this scientific approach. The right prefrontal cortex looks for errors in our hypothesis ("The saber-toothed tiger is not harmless"), and an adjoining region tells us to change behavior ("Run!").

- We can recognize and imitate behavior because of "mirror neurons" scattered across the brain.

- Some parts of our adult brains stay as malleable as a baby's so that we can create neurons and learn new things throughout our lives.

Brain Rules

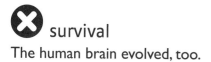 survival
The human brain evolved, too.

 memory
Repeat to remember.

 exercise
Exercise boosts brain power.

sensory integration
Stimulate more of the senses.

sleep
Sleep well, think well.

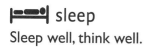 vision
Vision trumps all other senses.

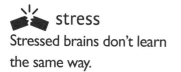 stress
Stressed brains don't learn
the same way.

music
Study or listen
to boost cognition.

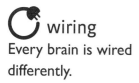 wiring
Every brain is wired
differently.

gender
Male and female brains
are different.

 attention
We don't pay attention
to boring things.

 exploration
We are powerful
and natural explorers.

Extensive, notated references
at *www.brainrules.net/references*

Acknowledgments

IN A LIST OF just about anything, items at the beginning and end are the easiest for the brain to retrieve. It's called serial position effect, and I mention it because I am about to list some of the many people who helped bring this project to fruition. There obviously will be a first person and a last person and lots of people in between. This is not because I see these folks in a hierarchy of values; it is simply because written languages are necessarily, cursedly linear. Please pay attention, dear reader, to the folks in the middle as well as to those at the end points. As I have often mentioned to graduate students, there is great value in the middle of most U-shaped curves.

First, I thank my publisher at Pear Press, Mark Pearson, the guiding hand of this project and easily the wisest, oldest young man with whom I have ever had the joy to work. It was a pleasure to work with editor Tracy Cutchlow, who with patience, laughter, and extraordinary thoughtfulness, taught me how to write.

Special thanks to Dan Storm and Eric Chudler for providing invaluable scientific comments and expertise.

I am grateful to friends on this journey with me: Lee Huntsman, for hours of patient listening and friendship for almost 20 years. Dennis Weibling, for believing in me and giving me such freedom to sow seeds. Paul Lange, whose curiosity and insights are still so vibrant after all these years (not bad for a "plumber"!). Bruce Hosford, for deep friendship, one of the most can-do people I have ever met.

Thanks to Paul Yager, and my friends in the department of bio-engineering at the University of Washington School of Medicine, for giving me opportunity. I'm also grateful to my colleagues at Seattle Pacific University: Frank Kline, Rick Eigenbrod, and Bill Rowley, for a spirit of adventure and for tolerance. Don Nielsen, who knew without a doubt that education really was about brain development. Julia Calhoun, who reigns as the premier example of emotional greatness. Alden Jones, amazing as you are, without whom none of my professional life would work.

And my deepest thanks to my beloved wife, Kari, who continually reminds me that love is the thing that makes you smile, even when you are tired. You, dear, are one in a million.

About the author

DR. JOHN J. MEDINA is a developmental molecular biologist focused on the genes involved in human brain development and the genetics of psychiatric disorders. He has spent most of his professional life as a private research consultant, working primarily in the bio-technology and pharmaceutical industries on research related to mental health. Medina holds an affiliate faculty appointment at the University of Washington School of Medicine, in its Department of Bioengineering.

Medina was the founding director of two brain research institutes: the Brain Center for Applied Learning Research, at Seattle Pacific University, and the Talaris Research Institute, a nonprofit organization originally focused on how infants encode and process information.

In 2004, Medina was appointed to the rank of affiliate scholar at the National Academy of Engineering. He has been named Outstanding Faculty of the Year at the College of Engineering at the University of Washington; the Merrill Dow/Continuing Medical Education National Teacher of the Year; and, twice, the Bioengineering Student Association Teacher of the Year. Medina has been a consultant to the Education Commission of the States and a regular speaker on the relationship between neurology and education.

Medina's books include *Brain Rules for Baby: How to Raise a Smart and Happy Child from Zero to Five*; *The Genetic Inferno*; *The Clock of Ages*; *Depression: How It Happens, How It's Healed*; *What You Need to Know About Alzheimer's*; *The Outer Limits of Life*; *Uncovering the Mystery of AIDS*; and *Of Serotonin, Dopamine and Antipsychotic Medications*.

Medina has a lifelong fascination in how the mind reacts to and organizes information. As a husband and as a father of two boys, he has an interest in how the brain sciences might influence the way we teach our children. In addition to his research, consulting, and teaching, Medina speaks often to public officials, business and medical professionals, school boards, and nonprofit leaders.

A

I

IBM, 114

Identical twins, brain wiring in, 94–95

"I'll Be Home for Christmas," 202

Imitation, 251
    newborns in, 247–248

Immediate memory, impact of sleep loss on, 48

Immune response, stress and, 64–65

Immune system, stress and, 72–73

Infant Behavior Questionnaire (IBQ), 214

Infants. *See also* Children
    acquisition of information by, 246
    brains in, 92
    cognitive abilities of, 76
    effects of music on premature, 220
    hypothesis testing by, 247
    music lessons for, 214–215
    object analysis by, 248–249
    object permanence and, 249
    reaction to new information, 250–251
    revelation of brain secrets by, 251–252
    testing of new information by, 246–251
    tongue testing and, 247–248
    use of visual cues by, 193–194

Information
    auditory, 142
    deconstruction of, 190
    encoding of, 106
    exploration in acquisition of, 246

meaning of, 139
    oral, 192
    pictorial, 192, 196–197
    processing of newly acquired, 250–251
    sensory, 154, 156, 167
    testing of new, by infants, 246–251
    transfer of, 193
    types of, 192
    visual, 142, 190

Information purification, 113

Infrared eye-tracking technology, testing of effect of pictures on attention, 196–197

Inhibition, 90–91

Insel, Thomas, 231

Insomnia
    Fatal Familial, 40
    healthy, 48

Inspiration, sleep and, 49–50

Intellectual dexterity, 74

Intellectual variability, 98

Intelligence, 96

Interest, attention and, 108

International Federation of Competitive Eating, 27

Interpersonal intelligence, 96

Interpretative activity as top-down processing, 169

Intervals, spaced, 149

Intrapersonal intelligence, 96

Intrinsic alertness, 111

Introductions, compelling, 140–141

iPods in music therapy, 201–202

IQ
    music training and, 206
    size of, 1

IQ tests, 96, 99, 101

Sleep debt, 47
  accumulation of, 45
Sleep deprivation, 40, 46
  chronic, 76
  effects of, 52
  research on, 39–40
Sleep hormones, 53
Sleepiness, transient, 45–46
Sleep loss
  cognitive performance and, 47
  effect on body, 48–49
  executive function and, 55
  logical reasoning and, 55
  mood and, 55
  motor dexterity and, 55
  quantitative skills, 55
  working memory and, 55
Slow consolidation, 152–153
Slow-wave sleep, 51
Smell, 169–170
  in boosting memory, 173–175
  in brand differentiation, 177
  in evoking memory, 173–175, 179
  motivation and, 176
  recall and, 175
  at work, 177–178
Smithsonian's National Museum of Natural History, Human Origins Program at, 9
Social bonding, oxytocin and, 217
Social cooperation, 14
Social sensitivity, differences in, 236
Social skills, music lessons in improving, 211–215
Software
  animation, 196
  PowerPoint, 197

Sound, 170–171
Spaced intervals, 149
Spacing out repetitions, 150–151
Spangenberg, Eric, 176–177
Spatial contiguity principle, 175
Spatial intelligence, 96
Spatial perception, depression and, 67
Spatiotemporal reasoning, 206
Speech, link between music and, 209–210
Sperm, size of, 226
Squire, Larry, 136
SRY gene (sex-determining region Y gene), 227
Starbucks, 176
Status, gender in negotiating, 235–236
Stimulation detection, learning and, 175
Stimuli
  aversive, 61, 69
  multiple senses in detecting, 170–171
  universal emotional, 114
Storage in declarative memory, 128, 159
Strengthening exercises, 25
Stress, 59–81
  absenteeism and, 72–73
  acute, 63, 232
  allostatic load and, 69, 70, 72
  anxiety disorders and, 73
  biology of, 62, 63, 66
  blood pressure and, 62, 71–72
  brain-derived neurotrophic factor (BDNF) and, 68, 80
  Brain Rule in, 57

Also by John Medina

Raise a smart, happy child through age 5

What's the single most important thing you can do during pregnancy? How much TV is OK for a baby? What's the best way to handle temper tantrums? Scientists know.

Through fascinating and funny stories, Medina unravels how a child's brain develops. You will view your children—and how to raise them—in a whole new light.

www.brainrules.net
